ENVIRONMENT & SUSTAINABLE DEVELOPMENT

K.A. Rasure

ENVIRONMENT AND SUSTAINABLE DEVELOPMENT

VOLUME - II

SERIALS PUBLICATIONS
NEW DELHI (INDIA)

First Published - 2007

ISBN: 81-8387-083-X

Published by

SERIALS PUBLICATIONS
4830/24, Prahlad Street, Ansari Road
Darya Ganj, New Delhi-110002 (India)
Phone : 23245225. Fax : 91-11-23272135
E-mail: serials@satyam.net.in

CONTENTS

VOLUME - II

LIST OF CONTRIBUTORS

Bamwerinde W., Forestry Research Resources Institute of the National Agricultural Research Organization, P.O. Box 311, Kabale, Uganda.

B. Bashaasha, Department of Agricultural Economics and Agribusiness, Faculty of Agriculture, Makerere University, P.O. Box 7062, Kampala.

W. Ssembajjwe, Department of Forest Management, Faculty of Forestry and Nature Conservation, Makerere University, P.O. Box 7062, Kampala.

F. Place, World Agroforestry Centre, P.O. Box 3067700100, Nairobi, Kenya.

S. Rajendran, Professor and Head, Department of Economics, Periyar University, Salem-636011.

George Chackacherry, Scientist & Officer in Charge, Centre for Water Resources Development & Management (CWRDM), Sub Centre, Neyyattinkara, Thiruvananthapuram, Kerala–695121, India.

K. Madhava Chandran, Scientist, Centre for Water Resources Development & Management (CWRDM), CWRDM Headquarters, Kozhikode, Kerala, India.

K. Sivasubramaniyan, Assistant Professor, Madras Institute of Development Studies, Madras.

K.K. Bagchi, Professor, Deptt. of Economics, North Bengal University, Darjeeling (W.B.), Pin–734013.

Swajan Kumar Dey, Research Scholar, Deptt. of Economics, North Bengal University, Darjeeling (W.B.) Pin–734013.

K.P. Kumaran, Associate Professor (CESD), NIRD, Rajendra Nagar, Hyderabad-500030.

Rudra Prakash Pradhan, Faculty, Economics and Finance Group, Birla Institute of Technology & Science (BITS), Pilani, Rajasthan–333031, India.

Vasudev Sedam H., Reader, Department of Economics, Gulbarga University, Gulbarga-585106 Karnataka (India).

Mohammed Younus K., Research Scholar, Department of Economics Gulbarga University, Gulbarga-585106

I. Satya Sundaram, Agriculture & Rural Development, and Research Director in Economics, 12/354-I, Rustumbada, Machilipatnam, Andhra Pradesh.

Falendra K. Sudan, Reader, Department of Economics, University of Jammu, Jammu and Kashmir, India.

G.N. Nagaraja, Extension Education Unit, University of Agricultural Sciences, Hebbal, Bangalore–560024.

Rajeswari, K. Extension Education Unit, University of Agricultural Sciences, Hebbal, Bangalore–560024.

Sukanya, C., Extension Education Unit, University of Agricultural Sciences, Hebbal, Bangalore–560024.

C.S. Singhal, Associate Professor, CHRD, National Institute of Rural Development Rajendranagar, Hyderabad–500030.

K.S. Venkateshppa, Professor of Economis, Department of Economics, Govt. First Grade College, Devanahalli, Bangalore(R)

K.A. Rasure, Selection Grade Lecturer in Economics Nrupatunga Arts, Com. & Management College, Sedam, Distt. Gulbarga (Karnataka)

R. Sundareshan D., Dept. of Agricultural Economics, TNAU, Coimbatore-641003.

Jagannath Lenka, Reader in Economics, North Orissa University, Sriram Chandra Vihar Takatpur, Baripada, Mayurbhnaj-757003 (Orissa).

Minati Mallick, North Orissa University, Sriram Chandra Vihar Takatpur, Baripada, Mayurbhnaj-757003 (Orissa).

B. Eswar Rao Patnaik, Senior Reader & HOD (Economics), S.B.R.G. Womens College Berhampur, Orissa.

Santosh Patnaik, 4th Year Civil Engg., N.I.T., Bhopal, Madhya Pradesh.

B.M. Kumara, Lecturer in Economics, University of Mysore.

A. Marimuthu, Lecturer in Economics, Madurai Kamaraj University College, Aundipatti-625512 Theni (Dist. Tamil Nadu)

Channabasavanagouda P., Research Scholar, Department of Economics, Gulbarga University, Gulbarga–585106, Karnataka, India.

VOLUME - II

24

THE PUZZLE OF IDLE LAND IN THE DENSELY POPULATED KIGEZI HIGHLANDS OF SOUTWESTERN UGANDA

Bamwerinde W., B. Bashaasha,
W. Ssembajjwe and F. Place

BACKGROUND

The development of the agricultural sector in Kigezi highlands in South Western (SW) Uganda has involved progressively more intensive use of land resources for cropping and grazing leading to unsustainable adaptation and changes in land use. Restoration of soil productivity in Kigezi highlands had relied on natural fallow practices and subsequent natural resource regeneration. However, existence of large proportions of uncultivated land in such densely populated area is evidence that some of this may be unintentional fallows (AFRENA, 1999). What is observed appears contrary to the expectation of increased annual cultivation in land scarce areas as postulated by the theory of induced innovation (Boserup, 1965). Some authors have suggested that the increased area under long fallows could be a farmer's management strategy to improve natural resource management and particularly fertility regeneration (Lindblade *et. al.*, 1998). Others have argued that disuse is because plots are very heterogeneous with some being very unproductive in terms of crop yields despite effort to increase production. Another explanation is that poor land markets could be inhibiting farmers from acquiring land that they would put into use and disposing of land that they have abandoned. Despite the above explanations, the economic determinants, causes and extent of land use decisions by farmers are not well understood.

As a result, there are as yet no agreed upon interventions to address the issue. In order to understand the impact of human activity on the highland ecosystem, it is important to identify and estimate the factors that influence farmers' decisions to put land to particular usage. Therefore, the main objective of this study is to determine the present status and causes of land abandonment in Kabale District, and propose policy options for sustainable management of land resources.

The Kigezi highlands are a region of high agricultural potential but with a high population density (about 300 per km^2) and population growth of about 2.2% per annum (Ministry of Finance, Planning and Economic Development, (MFPED, 2004). The highlands lie in SW Uganda at an altitude of 1500-2700 meters above sea level. The rainfall pattern is bimodal and ranges from 1000 to 1500 mm per annum. The temperatures are moderate with mean minimum of 13^0 C and mean maximum of 23° C. In Kigezi, soils are generally volcanic and were inherently fertile although some parts have less fertile Ferrasols and Andosols (Djimbe and Hoekstra, 1987).

PREVIOUS WORK ON FARMERS LAND USE DECISIONS

In this research, the practice of leaving land to rest for longer periods than the farmer would have desired *(unintentional fallow)* locally known as "*Eibija*" is what is referred to as abandonment of land. This is opposed to *intentional fallow* that is commonly practiced and known to farmers locally known as "*Hinga raza.*" Land abandonment has been observed in several other countries including Botswana, Japan and Lebanon (Mmopelwa ,1998; Arntzen, 1984; Kashiwagi, 1998; Zurayk 1994).

Mmopelwa (1998) empirically examined the proportion and factors causing unintentional fallowing in South East Botswana. He used data from 100 randomly selected farmers and simple statistical estimations. The study revealed that 20% of farmers unintentionally fallowed (short-term or long-term), 1% intentionally fallowed and 79% did not fallow at all. The factors attributed to unintentional fallows were the lack of draught power (86%), labor (82%) and rainfall variability (100%). Based on these findings two aspects of fallowing were noted. The first aspect was intentional fallowing, in which farmers purposefully left their land uncultivated for it to regain soil fertility. The second aspect was that of unintentional fallowing whereby farmers involuntarily leave their land uncultivated due to biophysical, social and economic factors beyond their control.

Grisley and Mwesigwa (1995) investigated the socio economic factors that influence seasonal fallowing in Kigezi highlands. The study revealed that 76% of farmers reported some cropland in grass fallow. Households reported an average of 26% of cropland under fallow. Model estimates revealed that intercropping and farm size were highly associated with the land fallowing decision but distance to fields was not. They recommended use of capital-intensive technologies such as terracing, agroforestry and use of chemical fertilizer, though currently unavailable, to overcome the problem of land being idle for a long time.

An empirical study by Barbier (1998) used a bio-economic model to investigate the theory of induced innovation and land degradation in the intensively cultivated areas of West Africa. The results of this study show that farmers prefer extensification to intensification even if land is marginal and produces lower yields. Under population pressure, the model revealed that over time soil organic matter and soil nutrients become scarce and the cost of land degradation increases. However, if the population is constrained and continuously farms the same plots, the economic results and welfare deteriorates especially when existing markets and technologies for agricultural improvements are limited.

The changes in land use system in Kabale District in Kigezi highlands between 1945 and 1996 as well as the current farm management practices were investigated by Lindblade *et. al.* (1998) using transects conducted in 1945 as baseline for measuring land use changes. They observed significant increase in area under fallow since 1945. For example, 19% of plots were left to fallow in 1945 as compared to 32% in 1996. Furthermore, the length of the fields rested had extended significantly with 95% left to fallow for more than half a year in 1996 compared to only 5% in 1945. Households with more than 11 plots reported 31% of their plots under fallow while farmers with less than 5 plots could fallow up to 25%. Finding more fallow to day compared to 1945 is perplexing given the perceptions of the local people. Rapid appraisal exercises conducted by various researchers reveal consistently that the perception of local people is that land fallow periods have decreased or are non-existent for some farmers. One reason could be due to definitions and assumptions: Lindblade *et. al.* (1998) considered all uncultivated plots as fallowed.

ANALYTICAL FRAMEWORK

The five land use options analyzed in this study include continuous cropping, short fallow, long fallow, woodlots and abandonment. At an individual farm level, decisions on whether to use a given plot of land for annual production or to put it under other land use options such as long fallow and woodlot is assumed to depend on the net present value of the option available to the farmer. For example, the practice of continuous cultivation with intermittent short duration fallows would only continue if normal returns on fixed capital would be obtained after all farm running costs have been paid. Continuous crop production and short fallow involves a modest capital investment, are short term in nature (0-1 year), and can result in optimal use of the plot if well managed (e.g. with adequate nutrient inputs). However, if land resources become depleted, potential future profits will be reduced until ultimately the plot is fallowed or put to some other land use option that would be less profitable in comparison to cropping if land was not degraded. Because the use of fertilizer is very low in Uganda as a whole, Kigezi no exception, the possibility that land become degraded with continuous use is real.

Fallowing is a land management strategy to intentionally increase soil productivity in the medium term (1-3 years). Consequently, land fallowing is an investment decision. This study argues and tests the proposition that not all farmers can afford this medium-term investment and therefore some of the plots that appear to be in fallow are in reality abandoned because of unintentional sub-optimal usage by farmers. The complexity between plot abandonment and long fallowing is that both practices appear to be indistinguishable by observation but based on different decisions. Therefore, long fallow, woodlots and abandonment decision by farmers are treated as different forms of long-term investment.

We assume that plot characteristics that affect productivity or the net returns will affect land use decisions. For example, plots of inherently poor soil fertility would be less likely to be put to into cultivation compared to other plots. The distance to the plot will also generate differences in labour costs and risks (e.g. of crop theft) and affect land use decisions. Household factors such as access to resource (e.g. labour, land) will also affect the net returns from the portfolio of plots and affect general land use decision-making at the farm level. As will be shown below, farms are characterized by extreme land fragmentation, with households owning 6 -10 plots widely scattered on the landscape. Thus, the possibility that a household will face widely varying net returns to an identical investment on their different plots is high. Farmers may also opt to abandon the plot after a short fallow. In some cases, farmers could revert land previously abandoned back to cropping depending on the availability of inputs.

ECONOMETRIC MODEL OF LAND USE DECISIONS

We estimate an econometric model of land use decisions using a snapshot or cross sectional data. The dependent variable (land use options) has five response levels representing the current land use option of a given plot. These are continuous cropping, abandonment, woodlot, long fallow and short fallow. The covariates include age of the household head, farm size, plot distance from homestead, household size, education level of the household head, gender of household head, polygamous or not, plot location, soil fertility status, presence of stones, plot size and slope. The logistic model is generally specified as follows;

$$Y_i^* = \beta' X_i + \varepsilon_i \tag{1}$$

Where, Y_i^* is the underlying latent variable that indexes the land use option on a given plot. X_i is a (kx1) vector of explanatory variables, β is a (kx1) vector of the parameters to be estimated, and ε_i is the stochastic error term.

A general expression (Greene, 2003) for the conditional probability for the five land use options (categories) is:

$$P(Y=j) \Big| X = \frac{e^{g_j(X)}}{\sum_{k=0}^{4} e^{g_k(X)}} \tag{2}$$

Where j is the category of land use. From this specification, the probabilities for each of the observed responses are easily derived.

EMPIRICAL MODEL

The model used to estimate the influence of plot and household characteristics on land use choices is specified as follows:

$$A= \beta_0 + \beta_1 FARMSIZE + \beta_2 HHSIZE + \beta_3 AGEHH + \beta_4 HHPMALE + \beta_5 HHFEMALE + \beta_6 EDUCPRIM + \beta_7 EDUCPOST + \beta_8 FERTMOD + \beta_9 FERTGOOD + \beta_{10} TOPMID + \beta_{11} TOPTOP + \beta_{12} STONFEW + \beta_{13} STONMANY + \beta_{14} SLOPMOD + \beta_{15} SLOPSTP + \beta_{16} PLOTSIZE + \beta_{17} PLODIS + e$$

Where;

A = Land use option (continuously cropped plots, short fallow, long fallow, abandonment and woodlots).

FARMSIZE	The total land holdings measured in hectares
HHSIZE	Number of household members measured in number of persons
AGEHH	Age of the farmer measured in years
HHFEMALE	Households headed by female (base case is monogamous households headed by males)
HHPMALE	Polygamous households measured as number of women living with the head.
EDUCPRIM	Household heads with education level of primary (base case is no formal education)
EDUCPOST	Household heads with education levels of post-primary or more
FERTGOOD	Terrace plots with good soil fertility
FERTMOD	Terrace plots with moderate soil fertility (base case is plots with poor soil fertility)
TOPMID	Terrace or plot is located in the mid slopes
TOPTOP	Terrace plots located on landscape (base case is valley bottoms)
STONFEW	Plot has few stones (base case is plots with no stones)
STONMANY	Plot or terraces has many stones
SLOPMOD	Plot or terrace is on moderate slope (base case is flat plots)
SLOPSTP	Plot or terrace is on steep slope
PLOTSIZE	The plot size in hectares
PLODIS	The distance of the plot from the home measured in Km

SELECTED VARIABLES FOR MODEL

Several exogenous or predetermined household variables are included in the model. The age of the household head is included because the probability of plot abandonment may differ with age. However, the relationship may move in different directions. Land markets exist, but are imperfect so older farmers may have had more opportunity to dispose of unwanted plots or have accumulated more wealth so as to avoid abandonment. On the other hand, they may have acquired many plots and found that they cannot adjust their land holdings as quickly as other factors like labor or capital have changed.

Household size, sex of household head, polygamous households, and education of household head are also included. Household size is hypothesized to have a negative relationship with abandonment. Female-headed households are hypothesized to be more prone to abandonment than male-headed households because of their difficulty in exercising rights or acquiring necessary factors of production. Educated farmers (especially those with post-primary education) are expected to have acquired knowledge in relation to improved methods of land management. They also have the ability to acquire loans to improve the land. However, they are also capable of diverting the resources to non-agricultural activities perceived to offer higher returns. This therefore may lead to less investment in soil improvement and increased land abandonment. Therefore, a priori the hypothesized sign is unclear.

The plot related variables included in the model were TOPTOP (plot on a hilltop) and TOPMID (plot on a mid-slope) with plots in the valley bottom acting as a base. Although both represented the topographic location of the plot, they are entered as separate variables because they are expected to influence abandonment of plots differently. It is expected that plots located on the hilltop are more likely to be abandoned compared to plots in the valley bottom because of poor accessibility. The plots in the mid slope are more likely to be either short fallowed or long fallowed because they are closer to the homes. The degree of soil fertility of a plot influences land use decisions. The variables FERTMOD (moderate soil fertility) and FERTGOOD (good soil fertility) are entered separately as dummy variables and plots with poor soil fertility acted as the base case. It is hypothesized that soil fertility is negatively related to abandonment. Therefore, both soil fertility variables are both expected to have negative signs.

The influence of plot stoniness on farmers' decision to abandon or cultivate the plots is also investigated. STONFEW (few stones) and STONMANY (many stones) are variables entered separately because many stones are hypothesized to encourage abandonment since it is difficult to open the soil (absence of stones is the base and omitted case). Absence of stones is expected to encourage continuous use of the plot. Therefore, the hypothesized sign is positive for STONMANY and STONFEW, but weaker in the latter case. The slope of the plot is also a characteristic expected to influence the use of the given plot. The slope variables are SLOPMOD (moderate slope) and SLOPSTP (steep slope). They are entered as separate variables and flat plots are used as base cases to verify the influence of slope on abandonment. The steeper the plot, the more difficult it is to cultivate and improve, hence greater slopes are expected to have a positive sign.

The hypothesized sign is positive for FARMSIZE and negative for PLOTSIZE. The larger the farm, the less likely it will be for a farmer to keep all of it in continuous cultivation, ceteris paribus. Larger plots have a negative sign and are expected to act as incentives for farmers to travel long distance and use the plots despite fragmentation. The variable (PLOTDIS) represents the distance from the homestead to the plot. It is hypothesized to have a positive sign. The further away the plot was from the home, the higher the probability of it being abandoned or long fallowed.

DATA AND SOURCES

To estimate the model, data were collected from 3 sub counties in Kabale district. Kabale district with a total land area of about 1850 km^2 has three counties and 19 sub counties. In each county, two sub counties were selected making a total of six sub counties. These included Kashambya, Kamwezi, Rubaya, Maziba, Bubale and Hamurwa sub counties. The unit of analysis in this study was a farmer's plot, but data were also collected on the farming household. Since factors related to plot abandonment were analyzed, a plot was used because it is the smallest unit of land on which cultivation takes place.

In order to allow for a wider coverage and representativeness, a random sample of respondents was used. Selection of sub-counties and parishes was undertaken using multistage simple random sampling. Twenty farmers were randomly chosen from each locality while 60 farmers were taken from each sub county. Interviews were carried out in 18 parishes. Data were collected from a total sample of about 360 households through use of a structured questionnaire and plot visits. Given the number of plots per household in the district, over 2,530 plots were covered in the survey.

EMPIRICAL RESULTS

Extent of Abandonment and Land Use Options

The current use of plots by farmers reveals that about 59.6% of the 2,530 plots are cultivated with crops. The short fallow and long fallow comprise of 8.1% and 13.7% respectively. The abandoned plots by farmers are about 10.1% of the total plots visited. The plots with woodlots comprise about 8.1%.

Farmers' reported that poor yields are the main reason for both long fallow and abandonment of fields. The survey reveals that of roughly one-quarter of plots either long fallowed or abandoned, 61% was due to soil fertility problems. Livestock grazing (10%), location of fields (11%), labor shortage by households (9%) and old age (2.3%) are also important factors for non-cultivation. Other reasons given include bush fires and farmers poor health that affected about 3% of the plots surveyed.

The results (Table 1) reveal that the average distance of plots from the homestead and roads is about 1.15 km and 1.5 km respectively ranging from 0 to 20 km. Because of the terrain, even when distances are small, it is difficult to reach the plots that look to be near homes. The mean distances to market from plots was 6 km. The average number of plots owned by the household head is 8 plots and ranges from 1 to 20 although scattered over the landscape. This scattering of land though sometimes used as a risk management strategy may partly lead to sub optimal usage of the plots.

Table 1
Variations in Factors that Influence Farmers' Land Use Decisions

Factors	*N*	*Minimum*	*Maximum*	*Mean*	*± S.E*
Distance of plot from home	2530	0.00	20.00	01.15	0.03
Total number of plots	2530	1.00	20.00	07.80	0.08
Farm size	2530	0.07	46.00	10.97	0.25
Age of household head	0360	18.0	82.00	45.12	0.27
Household size	0360	1.00	23.00	07.73	0.00
Size of plot	2530	0.03	25.00	01.26	0.04
Area under woodlot	2530	0.00	27.00	00.98	0.148
Area under short fallow	2530	0.00	25.00	00.68	0.119
Area under fallow	2530	0.00	26.00	01.23	0.149
Area abandoned	2530	0.00	57.00	01.07	0.193
Area under crops	2530	0.00	60.00	04.88	0.014

Note: Distance was measured in km, farm size and area in acres.

The mean household farm size is about 10 acres (4ha). This appears to be on the high side compared with existing data, which reports the average land holding as less than 1 hectare. Examination of data and the land use system suggest that an average land holding of less than 1 hectare per farmer in the Kabale district could be an underestimate. This appears to be based on the total land area divided by the total number of farm households, which does not cater for the land farmers own on the hillsides. Furthermore, unless one visits the plots, farmers tend to report fewer plots. Farmers tend to report only plots they use gainfully rather than plots lying idle.

The mean for household size is about 7 members per household with only 4 active members (55%) who are above 14 years. The rest are either elderly (0.5) or still young (3) members. Apparently, although this could be a good source of farm labor (especially the young) given the fact that universal primary education is in place, this figure leaves few members available for full time labor.

ESTIMATES OF DETERMINANTS OF FARMERS' DECISIONS

A single multinomial logit model was run, but because there are five outcomes, the estimates of the determinants of farmers' decisions for each separate category are given in Tables 2 and 3. The coefficients can be interpreted as the change in the log odds associated with a unit change in the respective independent variable. They represent the degree to which a one-unit change in an independent variable induces a change in likelihood from a cultivated plot to one of the other four land uses. Results of the four land uses are presented in the following order, short fallow model, long fallow, woodlot and abandonment models.

Short Fallow Model: The household level variables that are significant for short fallow were FARMSIZE and EDUCPOST. Other significant plot variables included, SLOPMID, FERTMOD, FERTGOOD, PLOTSIZE and PLOTDIS. The short fallow model results show that the odds ratio values for a given plot to be under short fallow are significant for EDUCPOST and FARMSIZE at a 1% level. For example, the odds ratio values obtained for EDUCPOST is 2.26. This implies that as the variable changes from no førmal education to EDUCPOST, the odds ratio for the plot to change from cultivated plot to be under short fallow increases more than 2 times compared to that of no formal education, the value of other independent variables held constant.

Table 2
Estimated Coefficients and Odds Ratio for Factors Influencing Farmers' Decisions
(Land Use for Short Fallow and Long Fallow)

	Short fallow		*Long fallow*	
Factor influencing farmers	*Coefficient estimate of the HH / plot factor*	*Odds ratio*	*Coefficient estimate of the HH / plot factor*	*Odds ratio*
Intercept	-1.799		-0.767	
AGEHH	0.003	1.002	0.013	1.013**
HHSIZE	0.003	1.003	-0.059	1.060
FARMSIZE	0.026	1.026***	0.005	1.006
EDUCPRIM	0.118	1.125	0.217	1.240
EDUCPOST	0.816	2.260***	-0.013	1.014
HHFEMALE	0.105	1.110	0.488	1.630**
HHPMALE	0.026	1.027	0.445	1.550**
TOPTOP	0.299	1.348	0.279	1.321
TOPMID	-0.723	2.060***	-0.526	1.692***
FERTMOD	-0.486	1.625**	-1.657	5.243***
FERTGOOD	-0.877	2.403***	-2.276	9.738***
STONFEW	-0.230	1.258	-0.172	1.188
STONMANY	-0.359	1.431	0.031	1.031
PLOTSIZE	-0.133	1.142**	-0.017	1.017
SLOPMOD	0.086	1.090	0.225	1.250
SLOPSTP	0.297	1.346	0.294	1.341***
PLOTDIS	0.126	1.134***	0.195	1.215

*, **, ***denotes statistical significance at the 10%, 5% and 1% level respectively.
HH denotes household

Table 2 Estimated coefficients and odds ratio for factors influencing farmers' decisions to either use land for short fallow or long fallow. Analysis was done using multinomial logit function against abandonment, woodlot, short fallow and long fallow based on continuous cultivation.

The plot variables such as TOPMID, PLOTDIST, and FERTGOOD significantly influence short fallowing. For example, the results show that the odds ratio of TOPMID was 2.06 and had a negative sign. This shows that the probability of a plot being under short fallow are 2 times less likely if a plot is located in the mid-slopes relative to the plots in the valley bottom. Other variables such as PLOTSIZE and FERTMOD are significant at the 5% level and have the hypothesized signs. The negative sign on plot size suggests that the larger the plots, the less likely they are put under short fallowed.

Long Fallow Model: Three variables related to household characteristics are significant at the 5% level. These include the age of the household head (AGE), households headed by females (HHFEMALE) and polygamous families (HHPMALE). Relationships between long fallow and age of household head have a positive sign. This implies that long fallow plots are more likely to belong to farmers who are older compared to those who are younger. The odds ratio value obtained for the variable (HHFEMALE) is 1.6. This shows that the probability (odd values) for a given plot to change from cultivated to long fallow is relatively higher for households headed by females compared to those headed by males. Furthermore, results show that the chance of long fallowing is greater for households with more than one resident wife compared to those with one resident wife.

The most significant variables are related to the plot. The significant differences are observed in soil fertility, location of the plot in the mid-slopes and plot distances at a 1% level of significance. For example, results show that the odds ratio for FERTMOD and FERTGOOD were 5.24 and 9.74 respectively and had the hypothesized negative sign. This implies that the odds ratio value of long fallowing for poor soils is 5 times higher compared to that for plots with moderate soil fertility and about 10 times greater than plots with good soil fertility. Again, the plots under long fallows are more likely to be located far from the homes compared to cultivated plots.

Woodlot Model: The woodlot model results (Table 3) reveal that one household variable (EDUCPOST) is important in determining the farmers' decision to establish woodlots and had the expected positive sign. Farmers with post primary education are more diversified into profitable livelihoods and could afford medium and long-term investments.

The plot variables that are significant at 1% level include, STONEMANY, FERTMOD, FERTGOOD, and SLOPESTP. Size of the plot is also significant although only at 10% level of significance. The fact that variables FERTMOD and FERTGOOD, reflecting plots with either moderate or good soil fertility are strongly negative and significant implies that soil fertility was one of the critical determinants for farmers' decision making to use plots for woodlots. Woodlots are more likely to be planted on plots with poor soils compared to fertile soils.

The stoniness is another plot variable that is highly significant. The odds ratio for stoniness in the woodlot model is 3.37. These results show that the odds for establishing woodlot increased to about 3.5 times among plots with many stones than the odds among plots with no stones. The steep slope variable is highly significant at 1% level. Farmers established woodlots on steeper plots as compared to gentler plots. The plot distance variable (PLOTDIS) represents the plot distance from homes. It had a positive sign but is significant only at the 8% level. The further the distance to the plots the more likely one observes a woodlot rather than a cultivated field.

Table 3 estimated coefficients and odds ratio for factors influencing farmers' decisions to either use land for Woodlot or abandon it. Analysis was done using multinomial logit function against abandonment, woodlot, short fallow and long fallow based on continuous cultivation.

Table 3
Estimated Coefficients and Odds Ratio for Factors Influencing Farmers' Decisions
(Land Use for Woodlot and Abandoned Plot)

Factor influencing farmers	*Woodlot*		*Abandoned plot*	
	Coefficient estimate of the HH / plot factor	*Odds ratio*	*Coefficient estimate of the HH / plot factor*	*Odds ratio*
AGEHH	0.011	1.012*	0.020	1.020***
HHSIZE	0.019	1.020	0.219	1.245
FARMSIZE	0.097	1.000	-0.006	1.007
EDUCPRIM	0.414	1.513*	0.781	2.184***
EDUCPOST	0.671	1.956**	-0.058	1.060**
HHFEMALE	-0.132	1.141	0.313	1.368
HHPMALE	-0.555	1.742*	-0.398	1.489
TOPTOP	-0.118	1.125	1.073	2.924***
TOPMID	0.243	1.275	0.255	1.290
FERTMOD	-1.950	7.029***	-2.279	9.767***
FERTGOOD	-2.892	18.029***	-2.669	14.425***
STONFEW	0.184	1.202	-0.316	1.372*
STONMANY	1.216	3.373***	0.419	1.520**
PLOTSIZE	0.083	1.087*	0.070	1.073
SLOPMOD	-0.030	1.031	0.739	2.094***
SLOPSTP	0.768	2.155***	0.522	1.685**
PLOTDIS	0.094	1.099*	0.211	1.235***

*, **, *** denotes statistical significance at the 10%, 5% and 1% level respectively.
HH denotes household

Abandoned Plot Model: The significant variables explaining abandonment include both household and plot level variables. Table 3 highlights two household variables that are significant at 1% level. These are age of the household head (AGE) and primary level education (EDUCPRIM). Post primary (EDUCPOST) is significant at the 5% level. The odds ratio value obtained for EDUCPRIM is 2.18. This implies that as the variable changes from no formal education to EDUCPRIM while the value of other independent variables are constant, the (odds) relative chances of a plot being abandoned increases by about 2 times. The sign reverses for post-primary education. Farmers with abandoned plots are more likely to have post primary education compared to those with cultivated plots. Farmers with post primary education may have greater financial resources with which to invest on their land, as shown in the woodlot results.

The plot level variables related to abandonment that are highly significant at 1% level include TOPTOP, FERTMOD, FERTGOOD, SLOPMOD and PLOTDIS from home. The coefficients for the variables moderate soil fertility (FERTMOD) and good soil fertility (FERTGOOD) have the hypothesized negative sign and are significant at a 1% level. The associated odds ratio imply that when the soil fertility changes from poor soils to FERTGOOD the chances (odds) of a plot being abandoned decreases by about 14 times other factors remaining constant. Furthermore, the odd ratio (probability) of a plot remaining under abandonment decreases more (10 times) as the plot changes from poor soil to moderate soil status.

The variable SLOPMOD had the odd ratio of 2.09. This implies that the (odds) probability for the plot to be abandoned are greater when the plot is located on the moderate slopes (about 2 times) compared to the odds when the same plot is located on gentle slopes. Presence of stones (many) and plots on steep slopes are significant at 5% level. Abandoned plots are more likely to be observed on hilltops and mid slopes compared to plots in the valley bottoms. Cultivated plots are more common in the valley bottom and abandoned plots on top of hills. The presence of stones is also another factor that influenced plot abandonment. The odd ratio value is 1.52. This implies that the odds for the abandoned plots are greater (1.5 times) where there is presence of many stones compared with the odds ratio for it where the plots has less stones.

CONCLUSION AND POLICY IMPLICATIONS

The results reveal that plot abandonment and long fallow is a common problem in the Kigezi highlands. Firstly, long fallow and abandoned plots comprises over 25% of the land area. Secondly, while plot characteristics are important in the long run, farmers' attitude, social and economic factors contribute to farmers land use decisions. The findings indicate that household characteristics such as post primary education and households headed by females significantly contribute to farmers' decision to put land under abandonment and long fallow.

What can be done to address the issue of abandonment or low investment depends partly on the specific characteristics of plot. Analysis of the long fallows and abandoned plots show that 26% and 40% of each face multiple constraints to cultivation (e.g. steep slopes and stony). Just under half of each type face one constraint. The remainder, 30% of long fallows and 15% of abandoned plots appear to be highly cultivable. But even here, further analysis found that between 60 – 70% of the cultivable plots are located at least 500 meters from the homestead. This will impinge upon incentives for intensification and investment in valuable or labour intensive systems. For the cultivable plots and nearby plots, there are already available technological options and they can be put into production if resources permit. The large number of plots with physical constraints implies that on the technological side, options that require relatively little labour or management, such as woodlots or fodder production, need to be identified.

There is a further group of uncultivated plots that are far from the homesteads of the current owner. The study reveals that distance to fields hinders productive use of these plots. As a result, policy makers and farmers need to consult and find meaningful ways to encourage free exchange of land, renting and selling of far off plots. This is feasible in the Kigezi highlands because there already exists an active land market. But the same land market has helped to create the high levels of fragmentation on the landscape. The process of improving the spatial distribution of plot holdings will be slow so long as only bi-lateral transactions occur. Thus, the government could play a role in facilitating interested exchangers of land to meet and enable multiple party exchanges to take place.

This study has some limitations. For example, the farm level data on which the findings and recommendations are made are mainly for one season. Data covering more years would be better suited to meeting the objectives of this study. These data would relate more closely to farmers production practices. Because of limitations of time and money, this study focused mainly on Kabale district. While this can be extrapolated for Kigezi highlands, this may not be true for other highland areas in the country because of varying social-economic and climatic conditions.

REFERENCES

AFRENA (1999), *Quarterly Report*, NARO/ICRAF Project.

Artzen, J.W. (1984), Changes in Rural Activities and Utilization of Natural Resources in the Periods 1979-1983: The Case of Malolwane, Kgatleng, District, National Institute of Development Research and Documentation, University of Botswana: 172 pp.

Barbier, B. (1998), Induced Innovation and Land Degradation. Results from Bio-Economic Model of a Village in West Africa, *Agricultural Economics*, 19 (1-2): 15-25.

Boserup, E. (1965), Conditions of Agricultural Growth, *The Economics of Agrarian Change under Population Pressure*. London, Allen and Uwin.

Djimbe, M. and Hoekstra, D.A. (1987), Agroforestry Potential for the Land Use Systems in the Bimodal Highlands of E. Africa. Nairobi, ICRAF.

Greene, W. H. (2003), *Econometric Analysis*, 2nd Edition, New York, Macmillan.

Grisley, W. and Mwesigwa, D. (1995), Socio Economic Determinants of Seasonal Fallowing: Smallholders in SW Uganda, *Journal of Environmental Management*, 42 : 81-89.

Kashiwagi, M. (1998), Reconstruction of Regional Agriculture, and Farmland Resource Management in the less Favored Areas under the "Staple Food Law", *Journal of Rural Economics*, 69 (2).

Lindblade, K.T. J., Carswell G., Tumuhairwe J. (1998), Mitigating the Relationship Between Population Growth and Land Degradation. Landuse Change and Farm Management in SW Uganda, *Ambio*, 27 (7)

Mmopelwa G. (1998), Factors Contributing to Land Fallowing in a Permanent Cultivation System: The Case of Semi Arid Botswana, *Journal of Arid Environment*, 40 (2) : 211-216.

MFPED (2004), Ministry of Finance, Planning, Economic and Development, Background to the Budget 2003/2004.

Zurayk R.A. (1994), Rehabilitating the Ancient Terraced Lands of Lebanon, *Journal of Soil and Water Conservation*, 49 (2) : 106-112.

25

ENVIRONMENTALY SUSTAINABLE ORGANIC FARMING SYSTEM: EMERGING ISSUES AND CHALLANGES

S. Rajendran

THE SETTING

After the introduction of modern farming techniques during 1960s along with expansion of irrigated area, the food grains production had increased substantially in India. At the same time it has been realized that due to intensive use of land with synthetic fertilizers and pesticides along with water guzzling crops soil fertility has been on the decline. In the early sixties itself it was suggested that unless judicious application of chemical inputs with local crop varieties were adopted the entire farming system will collapse[1]

Though literature on the organic farming is scarce, the findings of some researchers who have attempted to study the transformation process of conversion of a farm modern or traditional to an organic farm pertaining to western countries are available (see Lockeretz and others, 1984, Cacek and Langher, 1986, Andrew, 1987, Patriquin and others, 1986, Macrae and others, 1990 and Fukuoka, 1990). For tropical countries like India no comprehensive research had been done at farm level (Exception is Werf and Narayan, 1989 and Rajendran, 1998) and nevertheless a couple of descriptive studies (Rajendran, 2002 and Veeraragavathatham and Balakrishnamurthy, 2006) are available.

The population explosion influenced the government to make consistent efforts to increase food grains production by all means. In this process, as adequate care has not been taken to use the resources on a sustainable basis, the indiscriminate use of synthetic inputs rather 'chemicalistaion of farming' has increased over the last two decades and so have health hazards which are near disaster in some areas (Mencher 1991 and Pingali, 2001). The latest agrarian distress and farmers' suicides are also revealing the apathy of modern farming system (Narayanamoorthy, 2006 and Sathish, 2006).

In order to maintain soil health for sustainable agriculture growth the National Commission on Agriculture (GOI, 1976) called for a maximum utilization of organic matter[2]. This may be achieved by shifting to ecological farming. Arguments in favour of organic farming are the negative effects of modern farming system namely, capital intensive and chemicalistaion. Further, the organic farming system has other advantages like low water use. The export potential for organic farm products is bright as the western countries were eager to get pesticide free agriculture products due to heath problems caused by pesticides residues.

The paradox of the issue is that on the one hand the food grains production has to be increased to maintain food security and on the other, resources have to be managed and used on a sustainable basis. Furthermore, in order to attain social justice, equity and economic viability, sustainable resource management becomes crucial. Realizing these, there is a move to revive the traditional farming system, which the Indian farmers fostered and gained expertise from time immemorial.

Broadly the farming systems can be classified into three: Modern, Traditional and Organic. Modern agriculture makes intensive use of external inputs which are factory produced such as fertilizers and pesticides and knowhow for market–oriented production system. Traditional agriculture is a system whereby subsistence oriented farming is practiced using low level of both external and locally available inputs for subsistence. Organic agriculture aims to optimise the use of local resources by creating complex and diverse farming system and also seeks stability, growth and long lasting production level. Thus, of the three, organic farming system appears to be better option for sustainable agriculture development and in this paper an attempt is made to examine the suitability, issues and challenges relating to organic farming, which is considered as environmentally sustainable and economically viable in India.

METHODOLOGY

The present study is based on the information collected, by personal visits to 17 organic farms, covering various landholding sizes, adopting different cropping systems and practicing a number of complex farming techniques in diverse agro-climatic conditions in the country. Various central and state research centers and NGOs working on organic farming system have also been visited to elicit their views and experiences (Table 1). Most of the addresses were identified from the 'Green Farming' directory prepared by the Center for Science and environment, New Delhi and a few were by personal contact.

The non-availability of sufficient of farmers who practice organic farming, forced to adopt case study method, instead of survey technique. However, in the case study method there was a greater access to information as the researcher personally visited the farms plots, field crops, livestock and farm forestry. This method has also helped to gain more insights, into the diversity and complexity of various ecological farming techniques. The following section takes a look at the general characteristics (Table 2) of selected cases.

Table 1
Particulars of selected districts and farm

States	*Districts*	*No. of Organic Farms*	*No. of NGOs*	*No. of Research Centers*
Gujarat	Valsad	2	–	–
Maharashtra	Bombay, Dhane,Pune, Amaravathi, Sangali and Solapur			
Madhya Pradesh	Indore, Raisen & Hosangabad	5	–	1
Karnataka	Dharwad, Mysore & Bangalore	2	3	–
Tamil Nadu	Tirunelveli, Coimbatore, Pudukotrai, dindigul,	3	3	3
	Madurai and Thanjavur.	3	1	2
Kerala	Palakkadu	1	1	–
Pondicherry	Pondicherry	1	1	–
Total	21	17	9	6

Table 2
General Profile of Selected Farm Households

Farm sizes	*> 100 ha*	*25–100 ha*	*15–25 ha*	*Below 15 ha*
No. of farms	3	2	2	10
Level of education	Degree (2) & PUC (1)	Degree (2)	PUC (1) & Degree (1)	Degree (2), PUC (5), SSLC (3).
Major crops	Tea (1) Fruits (2)	Fruits (1) Cereals (1)	Fruits (2)	Mixed (3), Fruits(2), Cereals (3), Vegetable (2)
Farm Forest	2	1	1	5
Livestock Farm	1	2	2	9
Machinery	3	2	2	10
Irrigation	Borewell and Dugwell–12	Borewell and Dugwell–14	Borewell and Dugwell–3	Borewell and Dugwell–11
Access to Urban center	2	1	2	7

Note: Nos. indicate households and for irrigation number of pump sets possessed by the households.

Of the large farms, one was a tea estate owned by a larger tea trading company, located on the Western Ghats, other one was managed by the farmers on collective basis and third was owned by a single farmer. By and large all the farmers were educated above SSLC level and well informed with the research and development in agriculture and notably out of seven graduates two were graduates in agriculture science. This observation is unique in Indian agriculture is concerned as the Indian farmers are less educated and moreover there is a strong motion that educated (even they below to farming community) should enter into salaried occupation than farming. Some portion of the land had been kept for farm forestry to serve as biomass production and application and for other purposes.

On the forest land some farmers transplanted trees and in some cases they were allowed to grow naturally thereby a variety of plant species is grown. More number of livestock was observed among the farmers who operated less than 15 ha and the animal waste including manure and urine are used as organic matter. Machineries like tractors, tillers, pumpsets and sprayers were observed in almost all cases and one farmer did not possess any machinery. Accessibility with urban centers for both inputs and output was found good for all farms and a few of them had own transport facility to bring in inputs and dispose (carry) of farm products. In addition to these, majority of them (15) farmers possessed storehouses in the farms itself and in all the farms, commercial crops were raised with some exception. However, the cropping pattern was highly diversified, where the farmers grew a number of crops as mixed and intercropping and the cropping pattern was presented in Table 3.

Table 3
Cropping Pattern Adopted by Type Among Organic Farms

Fruits	*Vegetables*	*Tubers*	*Cereals*	*Others*
Pomegranate	Tomato	Turmeric	Paddy	Tea
Sopota	Cauliflower	Ginger	Ragi	Coffee
Guava	Carrot	and	Jowar	Cotton
Papaya ber	Beans	Tapioca	and	Coconut
Peach	Radish		Bajra	Mulberry
Mango	Guards			Sugarcane
Pineapple	Greens and			
Orange and	Onion			
Banana				

Of all these crops, cereals and vegetables were mainly cultivated for local consumption and organic growers prefer to use their products domestically as these are nutritious rich and free from chemical residues. Whenever surplus of these items was found then they were marketed and in a few cases it was not possible to grow some crops due to certain adverse agro-environment including non-suitability of soil condition.

Though organically produced fruits attracted takers from domestic market, the growers preferred to export as the price was quite high and among the 17 farmers studied, three farmers exported fruits, vegetables and tea abroad. One of the selected farms in Dindugul district in Tamil Nadu was the first in India to get its products certified as organic in 1987 and it was observed that fruits from the farm were not only organically produced but also processed with eco-friendly techniques like using solar energy[3]. Experts from the Institute of Marketing from Switzerland came here during harvest season for inspection to Dindugul farm and certification, which is not only uneconomical but also cumbersome especially to small organic growers.

Diversified farming system whereby a combination of trees, livestock and crops was required to effectively carry out organic farming system (Table 4). Under this system local species of trees, livestock and crop cultivars were preferred for their drought and pest resistant characters.

Table 4
Particulars of Resource Diversification

Type	Resources generated
Trees Legumes, creepers and Medical plants	Green manure, mulching, wind cover, biopesticides, fuel, fodder and timber.
Livestock Cattle, buffalo, sheep and Poultry	Manure, biogas, slurry and cattle urine
Crops Horticulture, cereals, Tubers, commercial and pulses	crops residues, fuel, copost Mulching, food items
Natural predators Owls, vultures, snake and Spiders	to check pests

Trees, creeps, legumes and medical plants were either planted or naturally allowed to grow on some portion of the farms and nursery for the seeding had been raised with local species on the individual farm itself. Shelterbelts raised on the borders of fields prevented banana and sugarcane crops from strong winds and soil erosion and trees attracted birds like owls and they controlled harmful pests on the field. Tree with large vegetative growth and canopy were regularly pruned to prevent shade, fell on the field crops and green and fallen leaves and twigs were used for mulching and composting. Biopesticides were prepared from the repellent plants with cattle urine to control pests and diseases.

Cattle manure was directly used for crop cultivation and in a few cases, slurry and sediments collected from biogas plants were used for field crops. In majority of the cases cattle urine was collected, stored and whenever pests occur the decoction was applied on the plants. In the sample none of the farmers reported that dung is used for fuel as they considered it as valuable manure. Goats were not preferred as they eat the nibbles of the plants and the locally produced fodder and feed were sufficient for the livestock and however, bovines and sheep were grazed on the fields and village commons.

Crop residues like cotton stake and banana leaves were used for cooking and mulching and this not only helped the useful organism to multiply but also preserved soil moisture by reducing evaporation and water holding capacity. Continuous mulching on the farms for five years, made the soil pores and organisms like earthworms naturally grew in abundant, which have provided valuable compost. Natural predators like owls, vultures and snakes, were freely allowed to breed on the farm, which effectively controlled pests and other unwanted creatures found. Many farmers opined that while natural predators ate some farm products, they saved considerable portion of them by killing rats and other harmful elements. Thus it is evident that use of locally available resources from livestock, trees, field crops and others had helped the farms to be self reliant for inputs.

PROCESS OF CONVERSION

Transition is the process of conversion of a farm from conventional or traditional (chemical farming system) to an ecological farming system (Werf 1990). Information on what made the selected farmers to switch over to organic farming system from modern farming was gathered and in majority of the cases the reasons were many and are presented on priority basis in Table 5.

All the selected farmers, except one, were practicing modern farming over the years and all the farmers belong to farming community by tradition who have gradually shifted some portion of their modern farm to organic farming system over the years. The process of transition took a minimum of 3 years to a maximum of 7 years, while 15 farms completely transformed, two farms partially transformed. According to the scientists and the farmers, the period of transition process depends upon the chemical contamination of the soil it had during the transition process. A couple of farmers took their soil and tested it from the laboratory to understand the chemical properties and soil fertility. Non-availability of scientific information including market channels and details on inputs for organic farming is a serious handicap for all the farmers Just by trial and error methods the farmers have been converting their farming system and some farmers used literature on sustainable and organic farming systems and as it was observed, literature which contain the farm techniques and others especially from vernacular language is not available.

Table 5
Factors for Transition

Sl. No.	*Factors*	*Responses*	
		Yes	*No*
1.	Low cost	17	–
2.	Health	17	–
3.	Incidentally	14	3
4.	Good market	13	4
5.	Environmentally sound	12	5
6.	Demonstration effect	12	5
7.	Fetches more price	11	6
8.	Taste and longer shelf life	10	7
9.	Increase in yield	8	9

Issues like low cost of various inputs needed for organic farming and health consciousness influenced all the selected farmers to shift from modern to organic farming. It is to be kept in view that by and large all the selected farmers produced food grains mainly for domestic consumption. Products from local crop varieties under organic farming system tasted well and farmers preferred and less than 50 per cent of the selected farmers reported that an increase in yield influenced them for transformation.

One farmer who graduated at Delhi University bought lands in an undulated terrain at Mysore district and successfully experimenting organic farming. Another two farmers mentioned that after having heard speeches and seminars on organic farming started transforming their farm. The above discussion shows that many selected farmers were largely influenced by economic, environment and other factors. Since the farming system among the selected farms is highly influenced by a number of complex factors, it is not possible to locate a specific aspect. Nevertheless, strong motivation from cost and environment led the farmers to transform their conventional farming system. In addition, it is as noted that the farmers were careful while transforming the farms thereby plot by plot choice instead of the whole farm was adopted during the transition process to avoid reduction in farm income. Unfortunately none of the neighbors of the selected organic growers came forward to switch over their farming systems, which needs to be probed indepth.

AGRONOMIC ACTIVITIES

The farm practices adopted in organic farms were complex and different from modern farm practices and a cursory look at them is necessary. In organic farming, land was not ploughed or tilled intensively instead light ploughing or rotating the topsoil with spade was encouraged. Because deep ploughing and furrowing, which were commonly done under modern farming, exposed the soil to heat, which in turn killed the beneficial organisms and water holding capacity was considerably reduced. According to the farmers light ploughing helped to prevent even water evaporation and nutrient loss and irrespective of cropping pattern this was followed.

Like any Indian peasant the selected organic growers should keen interest in seed and farm produced, healthy local seeds were carefully collected, processed, dried and stored for next crop season. Normally pots or grainages made from clay, bamboo splinters and twigs were used for storing seeds and lotus and *nochi* leaves were kept along with grains (seeds) to control pests like white moth, often found in seeds. While sowing seeds, pellets with dung and wet sand were prepared to prevent from extreme weather condition, birds and rats and this technique of sowing enhanced germination quality and yielded more as compared to normal sowing.

Normally flood irrigation was avoided to save water and prevent water logging under organic farming system and farmers were of the opinion that excessive use of water led to phenomenal weed growth. Modern irrigation techniques like drip system were practiced in some farms for horticulture and vegetable crops. Trenches were made in between rows for watering the fruit trees and in one of the cases slurry, dung and cattle urine were diluted and directly let into the paddy, napier grass and banana plots. Further research is required to find out, scientifically, the optimum level of manure application in the form of water and also the carrying capacity of the soil. During the survey the farmers reported that only 20 liters of water was used per plant (coconut and sapota) under trench method, whereas, 70 liters was used in drip system and 150 liters through flood system. In the present survey no farmer was found having access to canal or river water. Sixteen farmers possessed wells and tube wells and only one farmer relied on rainwater. Though a maximum of Rs. 50,000 had been invested for drip irrigation in one ha plot, water use and labour cost were considerably reduced (30 per cent) as mentioned by farmers. Two selected farmers dug contour bunds and smaller check dams which resulted in preventing soil erosion and increased water table in the neighbouring wells. Thus the survey reveals that watershed type of farming practices has helped the farmers to considerably reduce the water use thereby input cost.

Eco-friendly biofertilizer, urban and vermicompost were used as organic matter on the organic farms. Blue green algae, azolla, azospirillum, rhizobium and azotobacter were some of the biofertilizers commonly used and all these were efficient in fixing atmospheric nitrogen in the soil. Studies conducted at Tamil Nadu Rice Research Institute at Aduthurai revealed that inoculation of rhizobium and azospirillum adds 50-300 and 20-30 kg of nitrogen per ha respectively. While some of the farmers argued that if the soil health was maintained, the algae grew naturally, the scientists contented that unless the algae were screened scientifically there were possibilities to enter weed algae. Despite the fact that they are being cheaper–200 gm pocket was sold at Rs 2 and one kg was sufficient for one acre–the supply was far less than the demand as admitted by the officials in the agriculture department.

Municipal– supplied compost had greater demand among the farmers practicing organic agriculture who supplement it with compost and green manure. Due to heavy demand and lack of proper training in preparing compost, many a time, the farmers bought semi-decomposed manure with plastics, glass pieces, stone and other non-degradable items (Rajendran, 1993) which caused cut injuries to farm workers. In fact, municipal–prepared compost was an additional income to the municipalities and also kept the urban environment clean. It was estimated that on an average for one year, 66 lakh tones of urban compost and 2797 lakh tones of rural compost have been used for crop cultivation in India. The Compost Development Corporation set up in Karnataka was making all its efforts in this direction but not much headway had been made so far and sadly no other State is seriously attempting in this direction.

Earthworm casting improved the soil fertility and many pointed out that if the soil health was properly maintained by mulching, less tilling and stopping dreaded pesticides, earthworm multiple themselves. Bhidy's (1990) study indicated that income was higher to the tune of Rs. 2,000 per ha using vermicastings as compared to fertilizers and earthworm castings are rich in soil nutrient like nitrogen, sulphur and calcium. Nonetheless, researchers at the department of environment sciences in Tamil Nadu Agriculture University, Coimbatore called for a detailed research on earthworm species; the rate of their conversion of organic matter and reproduction patterns for commercial production and use.

Similar views were expressed by experienced farmers that earthworms, which effectively convert the soil, vermicastings have to be identified for multiplication, and grading has to be introduced to sell vermin compost. Since no grading was involved it became a lucrative business among the entrepreneurs said a scientist who was involved in earthworm research at the University of Agriculture Sciences, Bangalore. According to this researcher, three species such as *Eudrilus Engeniae, Perinyx arboricila and Eisenia fiteda* had been identified so far from 20 common species of India. But surprisingly in the market many species were available and since the demand was extremely high the price of one quintal of vermicasting varied from Rs 1,000 to Rs 4,000. Additionally, to avoid soil borne pathogens entering in the food, factory type compost preparation was suggested. It was sad to mention that while the public sector is less interested on this two private centers located in Pune have been doing detailed research on earthworm[4] and related issues.

Plant based repellents were common among the organic farms and farmers are keen on this. Plant leaves from *errukkan (Calotropis gigantica)*, mari gold, Aadathudai, Nochi, pongaamia *(Pongmia glabra), tulsi (Ocimum sanctum)*, and neem *(Azdirachta indica)* were used for preparing decoctions diluted with cattle urine. Farmers, possessed pucca cattle sheds collected cattle urine and stored in large containers or cement pits. While some farmers boiled the leaves, others directly soaked into water to use as decoction. In some cases it was observed that decoction was kept in pot or cement pit for one week, which had strong odour and later used as per repellent. Depending upon the season and crop type a few farmers sprayed it with sprayers and some used plant leaves to spray on the crops and arguably all these were done on trail and error basis and no scientific formula had been evolved on these aspects. Excessive usage of plant based biopesticides caused adverse effect on crop yields, damaging the colour of the fruits, when they were used on guava gardens as reported by one of the farmers[5] in the study area.

Pruning of selected plants have been reported from horticulture crops as such plants yielded more than the non-pruned plants. Crops like ber and guava were pruned before flowering for enhancing flowering and yielding. Observation from Solapur and Bangalore districts confirmed that pruning enhanced the number of shoots, in a plant, flowered more and led to increase in yield levels. Normally these crops yielded twice in year and pruning reduced production from two seasons to one season in a year but the yield level was more to offset the decline from two seasons.

Farmers collected tea residues from tea stalls to apply as manure. The weeds like congress grass (partheenium) and others grown on the fields were also collected and used as green manure. Except two farmers (Mysore and Bangalore) others did not pay much attention to prepare manure from cow dung and these farmers dug pits and cattle dung were deposited for decompose[6]. This created a conducive environment to proliferate beneficial organisms, in and around which were responsible for maintaining balanced nutrient in the organic matter and soil.

Invariably intercropping was practiced among the organic farms which not only kept away the pests and diseases but also generated additional resources and income. Castor plants raised on the boundaries of groundnut fields checked caterpillars and cultivation of sweet corn as an intercropping on the turmeric plots, kept at bay the pests like grass and leaf hoppers. Adoption of crop rotation, trapping, pruning infested plant parts and adjusting sowing/transplanting reduced the incidence and severity of insect pests and diseases in the organic farms.

Under the modern farming system integrated pest management (IPM) practice covering biological control, inter cropping and cultural practices were provided by the extension official. But lack of coordination among different wings within the agriculture department and among farmers IPM packages were not successfully carried out. Taking guidelines from IPM the selected farmers adopted a combination of environment-friendly biological and cultural techniques to control pests, diseases and weeds. From the above it is evident that the farm practices adopted under organic farming are farm based, simple, labour intensive and less cost to ultimately attain sustainable agriculture development.

COST AND RETURNS

Most of the farmers, though highly educated, did not maintained records of inputs cost, mainly because a lot of hidden costs were involved as the family members did certain farm operations and plant based products and manures were obtained from the farm itself. However, based on available data and information an analysis was done on the cost and returns of selected crops (Table 6).

Table 6
Distribution of per annum Cost Return (in Rs/ha.) from Selected Crops

Crops	*Total cost in Rs.*	*Gross returns in Rs.*	*Net Returns in Rs.*	*Net Returns over Cost (per rupee)*
Tomato	80,000	2,50,000	1,70,000	2.1
Tea	90,000	NA	NA	NA
Grapes	40,000	1,50,000	1,10,000	2.6
Paddy	10,000	30,000	20,000	2
Turmeric	7,800	30,000	22,000	2.9
Sapota	16,000	87,000	72,000	4.5
Curry Leave	NA	15,000	15,000	NA
Coconut	14,000	1,40,000	1,26,000	9
Groundnut	6,000	22,000	16,000	2.7

The input costs include labour days (both hired and owned), manure, water charges, land tax, seed value and others. In fact, cost for biopesticides was not taken into account as the leaves were collected either from the common forest or on the farmer's fields. Prevailing market wages rate had been taken into account to work out imputed value of family labour and imputed value for own manures and seeds was also considered. Water charges based on the local prevailing hire rate separately for diesel and electric motors were worked out. In the present exercise, the costs and returns were not averages of all farms and these pertained to respective crops from individual farm. Nonetheless, in order to examine the differences, per unit land (ha) costs and returns of various crops were worked out[7].

An overview of the and costs and returns for different crops reveals that per ha costs were more for tea[8] followed by the tomato and in that order and low inputs costs was observed for groundnut. On the other hand, leaving tea, tomato crop obtained more returns per unit of land followed by grapes and other crops. Crop wise performance was examined in relation to returns on investment and coconut found to be more remunerative followed by sapota, turmeric and grapes and the returns per rupee-invested were found relatively low on paddy.

The variation in costs and returns were due to differences in the variable costs and market imperfections. For instance, some inputs were available at lower rates and the market price for certain crop products was high, which together led to more returns for other selected crops. On the other hand, the farmers had to invest more for other selected crops and the market was unfavorable to sell such produces. These had led to lower returns for such crops. Moreover the selected farmers were located in different parts of the country and the market environment for inputs as well as output varied considerably. Nevertheless, the above analyses give an idea of the economics of organically grown crops across Indian farmers.

Collection of large quantity of green manure, fallen leaves and silt obviously required more labour than just applying mineral fertilizers, bought from the market. Similarly picking of repellent plant leaves and decoction preparation took more time and labour. In addition, application for these on the fields also required additional time and hence more labour was required in organic farms and hence organic farming system was found to be having wider scope to provide employment in the countryside.

As far as marketing of organically produced commodities was concerned the selected farmers did not come across much problem as only a handful of farmers had generated surplus and sold outside. This was mainly for the reason that people were willing to pay premium for organically produced commodities. Realizing the market potential for the organically produced items, outlets were mushrooming in major, urban centers like Mumbai, Bangalore, New Delhi and Chennai. However, the shopkeepers sold inorganically produced agricultural commodities as organically produced at exorbitant price after adulteration (Rajendran, 1994) and pathetically even scientists ad intellectuals have fallen prey to this as there is complete lapse on the part of grading and inspection. Some shopkeepers who bought directly from the growers, paid lower price and sold at higher rates thereby cheating both the producers and consumers. According to a selected farmer, who was also president of District Sapota growers association in Dhane in Maharashtra, only 50 per cent of the consumer's rupee went to the producers. As some of the agencies started exporting organically produced items, the farmers had greater demand for their produce and however, since the farmers were scattered spatially processing, packing and transportation problems cropped up. A mechanism has to be arranged to coordinate and manage this crisis as Maharashtra Grape Growers Association contemplated to establish its market globally.

Table 7
Organic Food Market–A Glance in 2000

Markets	*Retail sales (in billion US$)*	*Total food sales (in per cent)*	*Expected growth (in per cent)*
Germany	2.2–2.4	1.25–1.5	10–15
United Kingdom	1.0–1.05	1.0	25–30
Italy	1.0–1.05	1.0	15–20
France	0.75–0.80	1.0	15–20
Switzerland	0.42–0.45	2.0–2.5	15–20
Denmark	0.35–0.37	2.5–3.0	10–15
Austria	0.25–0.30	2.0	10–20
Netherlands	0.22–0.27	0.75–0.10	10–20
Sweden	0.12–0.15	1.0	10–25
Other European countries	0.30–0.40	NA	NA
Europe total	7.0	–	–
United States	8.0	1.5	15–20
Japan	2.50	NA	NA
Total	17.50	–	–

Note: Official trade data are not available. Compilations are based on rough estimates. The figures for Japan include 'green products'.
Source: Rajendran, 2002, p.111.

The organic food has good market in western countries as well (Table 7). In United Kingdom for example, the organic food market has grown from $ 8 million in 1985 to $ 80 million in 1990. This was projected at 4240 million in 1994 (Raj and others, 1996). Signing of GATT by 124 countries including India has opened up the global market for all member countries and export of organic farm products can certainly increase the forging earnings in a big way. Therefore, there is an urgent need to give emphasis for processing and grading to export for attracting global market besides supplying to domestic market.

SUMMARY AND CONTOURS OF CHALLENGES

The present paper highlighted various issues pertaining to environmentally sustainable organic farming by taking cases from different states along with available supportive evidence. All the farms covered in the present study possessed by large the resources required for farming and also most of the farmers had good network with the government machinery, NGOs and other fellow farmers. In fact, some even got wider publicity in the media and some farmers got international assistance, guidance and support and understandably all these gave moral support to the farmers to successfully venture into organic farming. Commitment was found to be the key factor in experimenting the organic farming system among the selected farmers and some farmers participated at national and international forum in disseminating various aspects of organic farming principles. Awards like Krishi Bhushan. Vasantha Rao Naik award and Bajaj award were presented to some farmers who were covered in the present study for their contributions in the area of sustainable organic agriculture system.

The resources used on the farms were collected locally either on the farm itself or in the vicinity including common property resources like village forests and tanks that served as base for collecting some resources required in farming. As a result of these, the cash component in the total farm expenses were found to be low and labour used was more.

Though commercialization of farming was not explicitly visible, products such as fruits, tea, vegetables, sugarcane, cotton were marketed and a major portion of cereals, pulses and oilseeds was consumed locally. If the farmers join hands on cooperative basis or any other mean on collective line there is a wider scope to attain greater economies of scale. Supply of inputs such as biofertiliser and vermicasting was far from satisfactory and even the government officials admitted that only 10 per cent of the total demand was met in the case of biofertilisesrs (Rajendran, 1995).

Some NGOs had successfully experimimeted organic farming system but not in a position to disseminate it to the farmers at large. The classic example was Friends Rural Center (FRC) an NGO at Hosangabad in Madhya Pradesh. This center with tireless efforts had proved that organic farming and natural farming were ideal for crops like paddy, soybean and mungbean. Famous book "One Straw Revolution" by an internationally known Natural Farmer, Fukuoka was first released in this center way back in 1990. Due to lack of leadership and paucity of funds, research was confined to four wall of the center and as a result, many committed and well-trained personal left this center and therefore genuine NGOs have to be identified and encouraged for experimenting and promoting organic farming system.

Other crucial issue is the research component. Both NGOs and the government research centers have been experimenting different components of organic farming system. Our experience is that some committed NGOs had taken up a holistic approach on this subject and even disseminated and demonstrated organic farming techniques in the training and fields. But the government centers have been carrying out compartmentalized research on this area. For instance, research on different pest repellent plants was done in different laboratories of the country. Though no packages had been evolved so far a few private companies started marketing biopesticideds (Jain, 1994).

The most important issues is that not scientific studies are available on this crucial subject. As pointed out by Nadkarni (1991) *"certainly, more research is needed to evolve suitable practices and to demonstrate the potential of natural farming in a variety of conditions*[9]. Since agriculture needs location specific techniques, the agriculture boffins have to take up research on variety of crops and demonstrate them across different agro-climatic conditions. Core issues like economic threshold levels of pest and disease attack and suitable eco-friendly remedial measures are to be found out by keeping technical feasibility and economic viability in addition to environment sustainability in view. Farmers close to the cases of present study were not well informed about organic farming practice. A cross section of such farmers reported as that "we get information on hybrid seed, chemical fertilizers and pesticides from the extension officials but no one explains about organic farming" and this issue has to be seriously viewed by the extension and NGOs workers.

Organic farming is economical, environmentally sound, socially adaptable and the invisible positive effects of it on the farm environment are many. For example, continues application of eco-friendly manures enhanced and enriched soil fertility and water use was scientific and the resources were stocked on the farms itself. All these helped to use the crucial resources on a more sustainable basis and to increase in yields that will sustain the farm economy in the long run.

The union Ministry of Agriculture (UMA) announced in 1994 a package of progarammes on organic farming system at the cost Rs. 26 crores. Technical committee set up by the UMA had recomeended this to maintain soil health and production to meet the increasing demand for organically produced food. Under this, scientists, farmers, extension officials and NGO representatives undergo training on organic farming system and however this has not yielded any positive result yet. In western countries like UK and Germany incentives are given to the farmers who are interested to undertake organic farming. Under the five year scheme, the UK government pays $ 104.16 per ha to the farmers who covert to organic methods initially reducing this $ 37.2 per ha in the last year (DTE, 1993).

Nonetheless, opinion among the officials in UMA and Indian Council For Agriculture Research ICAR) were divided on the viability of organic farming system. While some were seeking a trial on it through ICAR research centers, a few were skeptical about this[10]. The contention of the later group is that switching over to organic farming from modern farming would shake the food security of the country.

The pertinent question is that how much time a farm is required to transform from modern to organic farming system. It depends on the intensity of fertilizers and pesticides used on a given soil and normally the transition periods range from (3 to 7 years). This period is not only to transform but also to maintain and stabilize the production levels as in the initial years of transformation, the production declines considerably. A solution to this complex issue is, as the farmers suggested, a portion of farm land, may be one third, could be put under transformation process every year till entire farm is switched over to organic farming. No farmer could able to assess the resource base required for carrying out organic farming on a economically sustained basis. But based on the experiences in majority of the cases the farmers had sugesested that one ha of land with cattle heard is deficient to feed 5-member family. However, as some experienced farmers pointed out that type of land (irrigation/dry), inputs, resource endowment differ from region to region it needs a comprehensive and detailed research on the whole question of sustainable organic agriculture system. Therefore, any policy towards organic farming system has to consider various issues like sustainability, economic viability, investable capacity, employment and marketing.

NOTES

1. M.S. Swaminathan the brain behind the green revolution, cautioned way back in 1968 as "Intensive cultivation of land without conservation of soil fertility and soil structure would lead ... us into an era of agriculture disaster in the long run, rather that to an era of agriculture prosperity".
2. The NCA report mentioned that "there should be greater application of organic matters and wastes in order to permanently improve the physical and biological conditions of the soil for which recycling of processed waste in all available forms is a matter of urgent necessity"(P 101).
3. By a solar hot air collector fruits were hygienically dehydrated and chemical, preservatives and colouring agents were not used while processing.
4. According to a Pune based earthworm research institute, broadly there are two types of worms deep furrowing and surface worms. Some reported that surface worms were preferred as they consume all types of garbage and multiply quickly. Other group suggested that deep furrowing worms were preferred as they survive in stress condition and make the soil more pores.
5. Madras based agriculture consultant A Sankaram during our personal discussions cautioned that unwanted proportion of application of materials on the soil and plants is likely to adversely affect the plant growth and yield and hence a comprehensive research is sought from the research agencies.
6. In India there are two well knowing methods, NADEP and Indore to prepare compost. Under these methods local plant materials and sand are used and dung is covered with mud. Within 3 months time the well-decomposed manure is obtained. Some state governments provide financial assistances for this; nevertheless, practically many farmers do not follow these in India.
7. In many cases, the fields were cultivated with mixed crops including horticulture crops and hence, it was difficult to work out the costs and returns

for all crops grown in a particular plot. Nonetheless, costs and returns were estimated for a major crop grown in a given plot.

8. Our consistent attempts to get market price for organic tea from the selected estate proved futiie. The reason given by official in the estates was that since organic tea is entering first time in the market, it was decided to maintain the market rates secrete to avoid competition.
9. After having visited one of the organic farms (which is also one of the farms in the present study) Nadakarni (1991) said "there is definitely a way out of the crisis of increasing costs facing Indian, agriculture".
10. Personal discussions with ICAR's assistance director generals and New Delhi revealed that pure organic farming techniques might reduce production in the initial stage. According to them, to achieve the production food grains chemical fertilizers are a must. Similar views were expressed by Reddy (1995), as "in the prevailing conditions in developing countries, organic farming are not only unadaptable at the farmer level due to low profitability but also at the policy level consequent to the long gestation periods require for this system to be as productive as the conventional system (fertilizer intensive). M.S Swaminathan, during the personal interview cautioned that external inputs like fertilizers have to be judiciously used with eco-friendly inputs like biofertlisers to maintain production; levels, otherwise the food security might be weaken.

REFERENCES

Andrew J. (1987), Making the Transition to Low Input Agriculture: A Farmer's Perspective, *American Journal of Alternative Agriculture,* 1 (3), pp. 37-38.

Ann. (1994), Green farming (Directory), *Center for Science and Environment,* New Delhi.

Ann (1993), Incentives for Organic Farming, *Down to Earth,* December 31, p. 21.

Ann. (1993), Realizing the Virtues of Organic Farming Network, *Down to Earth,* November 30, p. 49.

Bhidy M.R. (1990), Organic Agriculture System – A Holistic Method, *Institute of Natural Organic Agriculture*, Monograph, Pune.

Cacek Terry and Langher L. Linda (1986), The Economic Implications of Organic Farming, *American Journal of Alternative Agriculture,* 1(1): 25-29.

Fukuoka M. (1990), The One–Straw Revolution, *Friends Rrural Center,* Rasulia.

Government of India (1976), Abridged Report of the National Commission on Agriculture, *Department of Agriculture and Co-operation,* New Delhi.

Jain Rimjhim (1994a), Biopesticides–Boost to Eco–friendly farming, *Down to Earth,* 2 (19), p. 9.

Lockeretz E. et.al (1984), Comparison of Organic and Conventional Farming in the Corn Belt: Current Technology and its Role in a Sustainable Agriculture, *Maidon ASA*, pp. 37-48.

Macrae R. et al (1990), Farm Scale Agronomic and Economic Conversion from Conventional to Sustainable Agriculture, *Advance in Agriculture,* 43, pp. 155-198.

Mencher Joan P. (1991), Agriculture Labour and Pesticides in Rice Regional of India: Some Health Considerations, *Economic and Political Weekly,* XXVI (39): 263-8.

Nadkarni M.V. (1991), The Crisis of Increasing Costs in Agriculture: Is There A Way Out? In India: The Emerging Challenges, *Ed. Nadkarni M.V, A.S. Seetharamu and Abdul Aziz, Sage,* New Delhi, pp.153-174.

Narayanamoorthy N. (2006), State of India's Farmers, *Economic and Political Weekly*, XLI (6) : 471-3.

Patriquin D. et al (1986), Observation on a mixed farm during the Transition to Biological Husbandry, *Biological Agriculture and Horticulture,* 4, pp. 69–154.

Pingali P.L. (2001), Environmental Consequences of Agricultural Commercialization in Asia, *Environment and Development Economics,* 6 (Part 4), October, pp. 483-502.

Raj Harender, M.L. Bhardwaj and N.K. Sharma (1996), Need For Eco-friendly Farming, *Employment News,* 10-16, p. 1.

Rajendran S. (2002), Economic and Environmental Dimensions of Organic Rice Cultivation in South India paper presented at the 1st *RDA/ARNOA International Conference on Basic Standard for Organic Rice Cultivation,* held from November 12 to 15, at Suwon/Cheonan, Korea.

——— (1998), An Explanatory Study on the Organic Farming System in India in *Living Resources for the Millennium 2000,* Ed. S.J. William, Loyola College, Chennai.

——— (1995), Eco-friendly Biofertlisers–Need of the Hour, *The Hindu Business Line,* February 22, Madras.

——— (1994), Too Little of a Good Thing–The Increasing Demand for Organic food is causing Shortage and Adulteration, *Down to Earth,* 2 (220, pp. 22-23.

——— (1993), Organic Manure is a Source of Income for Civil Bodies in Tamil Nadu, *Down to Earth,* December 13, p. 23.

Reddy Ratna (1995), Environment and Sustainable Agriculture Development: Conflicts and Contradictions, *Economic and Political Weekly,* XXX (12): 21-7.

Sathish. P. (2006), Institutional Credit, Indebtedness and Suicides in Punjab, *Economic and Political Weekly,* XLI (26): 2754-2761.

Swaminathan M.S. (1968), Keynote address in the Indian Science Congress held (January) at Varanasi.

Veeraragavathatham D. and G. Balakrishnamurthy (2006), National Seminar on Convergence of Technologies for Organic Horticulture, *Abstracts of the papers,* July 20-21st, Ed. Tamil Nadu Agricultural University, Coimbatore.

Venkataramani G. (1991), Sustainable Agriculture–Intensive Research will pay Off, *The Hindu Survey of Indian Agriculture,* Chennai, pp. 17-19.

Werf E. van der and B. Narayan (1989), A Socio-economic Study of Ecological Agriculture in South India, *AME and ICSIM,* Bangalore.

26

IMPROVING IRRIGATION EFFICIENCY IN KERALA THROUGH PARTICIPATORY IRRIGATION MANAGEMENT: AN ANALYSIS

George Chackacherry & K. Madhava Chandran

BACKGROUND

The conventional understanding of the irrigation sector was primarily limited to design, construction, operation and maintenance of irrigation systems. However, the gap existing between irrigation potential created and utilized in different parts of the world forced governments to look for suitable alternatives. Accordingly, the concept of irrigation management was recognized. This perspective added several non-physical functions such as planning, monitoring, financing and evaluation to the realm of modernization of irrigation systems. Irrigation is not only a physical concept, but also one involving a much broader set of management issues (World Bank, 1998). The International meeting on modernization of irrigation system operations, held at Aurangabad in 1998, defined irrigation modernization as a process of technical and managerial upgrading of irrigation systems (as opposed to mere rehabilitation), combined with institutional reforms with the objective of improving resource mobilization and water delivery service to farms. Countries, which once promoted more government involvement in irrigation projects, are now adopting policies that create incentives for users to take over management of operation and maintenance of the projects, while government agencies focus on improvements at the main system level.

PARTICIPATORY IRRIGATION MANAGEMENT

Participatory Irrigation Management (PIM), also known as Irrigation Management Transfer (IMT), is an attempt to increase direct involvement of farmers in irrigation management. It implies expansion of the role of farmers and the private sector and reduction of the role of government in irrigation management through transfer of authority and responsibilities, either in part or full, to farmer organizations. Such reforms are intended to improve irrigation performance by involving farmers, who are the most important stakeholders. PIM would lead to some form of joint management of irrigation systems, with the State responsible for more tasks at the higher level of projects (main system) and farmer organizations for tasks at lower levels (branch canal/ distributory/field channel). PIM is the keystone of water resources modernization reforms being implemented in about 60 countries around the world. Efforts to adopt PIM are often initiated by governments mainly due to shortage of funds to manage irrigation projects, which is also an outcome of factors like poor recovery of water charges, high overhead costs etc. It is expected to reduce operation and maintenance costs for governments, while enhancing the efficiency and sustainability of irrigation projects. In addition, the policy orientation of funding agencies like World Bank, and the shift towards decentralization/ privatization also contribute towards adoption of measures such as PIM.

The Ninth Five Year Plan of the Government of India had laid thrust on promoting PIM (MOWR, 1996). At the national level, efforts to initiate farmer participation started during the 1980's with the establishment of Command Area Development Authority (CADA) under various irrigation projects. Improvement in irrigation management and crop productivity through promotion of participation through registered farmer organizations was the main objective of this programme. This participatory approach was envisaged as a forerunner to PIM, to be implemented at a later stage under the irrigation projects. CADA was established in Kerala during 1985. A three-tier system, with outlet-based farmer associations at the base level, canal committees at the middle level and project committees at the upper level were envisaged under CADA. According to the report of CADA (2004), 4011 farmer associations were formed in 16 irrigation project commands in the State. However, studies have shown that only about 30% of them are functional, while the extent of farmer participation in the functioning associations is also low (Chackacherry, 1995; CWRDM, 1999). However, CADA programme has been able to provide a platform for a change in the outlook of farmers towards a decentralized and democratic system of management.

Kerala has no rich tradition in the management of irrigation systems by farmers, though farmer managed irrigation systems exist in various parts of the country, for hundreds of years. This may be attributed to better availability of water resources in Kerala in the past, which might not have encouraged community action for irrigation management (Chackacherry, 1995). Serious initiatives had not been taken up in Kerala for irrigation management transfer to farmers through PIM, especially in the major/ medium irrigation sector. However, neighbouring States like Tamil Nadu and Andhra Pradesh are well ahead with regard to implementation of PIM. On the one hand, under the present set up in our State, no part of the irrigation system can be handed over for management to farmer organizations, since they are not socially capable of taking up the tasks assigned to them; they should be either reorganized/restructured or adequately strengthened. On the other hand, the government agency concerned with these activities is neither physically nor socially equipped for taking up joint management with farmers. There are several problems that may hinder the implementation of PIM in Kerala such as less number of full-time farmers, small sized holdings, lack of sufficient interest among the farmers, reluctance of officials, etc. It was in this background that a study was undertaken jointly by the Centre for Water Resources Development & Management (CWRDM), Kerala and CADA of the State, with the main objectives of suggesting measures for reorganizing/ restructuring the existing farmer associations for implementing PIM in Kerala, and to suggest improvements in the agency set up for supporting the process.

METHODOLOGY

The study was carried out in three irrigation projects in Kerala, namely, Neyyar, Chalakkudy and Malampuzha. Primary data was collected through Participatory Rural Appraisal (PRA), questionnaire and interview method from farmers and officials. Farmer associations were selected randomly from the list of live associations maintained by CADA. PRA was carried out with 41 farmer associations under the irrigation projects. The tools used for PRA include focus group discussion, semi-structured interview and pair-wise ranking. 650 farmers were consulted. Discussions were held with canal committee chairman, canal committee members and project committee members. 30 women randomly selected from ten farmers' association were consulted regarding their involvement in irrigated agriculture, following semi-structured interview method.

Besides farmers, 300 officials of the Irrigation department, CADA and Agriculture department from Lascar to Chief Engineer level were covered during group discussions and interviews. Their hopes and apprehensions about PIM were assessed, and opinions and suggestions were pooled. Members of Legislative assembly, leaders of political parties, Secretaries to Government etc. were also interviewed under the study. Secondary data from earlier studies carried out by the investigators, reports of Ministry of Water Resources etc. were also made use of in the study.

FINDINGS AND DISCUSSION

The responses obtained from farmers and officials have been analyzed and interpreted as factors to be considered/pre-requisites for implementing PIM in Kerala, as outlined below (CWRDM & CADA, 2001).

1. **Favourable Orientation of Politicians and Administrators:** At the outset, it is necessary for the political, administrative and irrigation agency leadership to take interest in adopting PIM. Andhra Pradesh pioneered in implementing PIM mainly because of the strong will and commitment of the political and administrative machinery there, including the Chief Minister, which resulted in the enactment of *The Andhra Pradesh Farmers Management of Irrigation System Act,* 1997.

The present study has also shown a positive orientation on the part of the administrative and irrigation agency in Kerala towards PIM.

2. **Changes in Organizational Set up:** Irrigation agencies (CADA and Water Resources Department) and farmer associations constitute the present organizational structure under irrigation projects. This structure is primarily designed for full agency management, and there is no real opportunity for farmer or joint management. FAO (1995) mention that the major functions under irrigation projects such as governance, financial management, and execution mode and accountability pattern are all adapted to agency management only. According to CWC (1992), the task of organizational and procedural changes is to re-focus the emphasis of irrigation management from serving the macro level bureaucratic priorities to the dynamic field conditions and micro demand for services. Hence, changes are required in organizational structure with regard to clear water rights at the system and farmer levels, clearly designed line of authority between farmers and agency, and supportive accountability and incentive systems for management for implementing PIM in Kerala. Farmers should also be equipped with new managerial and technical skills related to irrigation management. Farmers, who were initially treated as beneficiaries under CADA, should become partners in planning, implementation and evaluation of irrigation development programme, while implementing PIM. The irrigation agency must be made multi-disciplinary, and the bureaucracy should back up efforts for management transfer and assign specific roles to officials to facilitate this. Farmers should have more opportunities for communication and interaction with officials. Sufficient number of awareness/ training programme should be organized for farmers on various aspects of PIM. Systems for regular monitoring/evaluation of PIM should also be instituted.

   The process of management transfer to farmers from the irrigation agency requires setting up of farmer organizations, which are capable of assuming responsibilities. Reorganization of farmer associations existing under CADA, and providing them statutory powers would be very helpful in ensuring effectiveness under PIM. Community organizers will have to be employed to help farmers in establishing Water User Associations (WUAs). NGO's can also offer valuable service on this aspect. For WUAs to be socially and economically viable, a minimum area of operation has to be decided, so that the cost of management does not become too high. However, the area should not be too big, in which case, the number of farmers will become unmanageable. One WUA can be made to serve an area of 200 to 300 ha, depending on hydraulic boundaries.

   Water User associations (WUAs) will have to enter into partnership with the irrigation agency and perform functions such as procurement, allocation and distribution of water, generating contributions from members in cash and kind (labour) for maintenance works, checking indiscipline, settling disputes and overall supervision and control of the irrigation system handed over to them under PIM. At the same time, the irrigation agency should be in a position to deliver the agreed quantities of water at the right time and place. Then only, farmers will feel motivated to carry out the tasks assigned to them under PIM. A study carried out by CWRDM in Kerala has revealed that the cardinal factor influencing farmer participation under CAD programme is timely, adequate and equitable water availability (Madhava Chandran and Chackacherry, 2004).

Accountability of WUAs to all the members is an important principle for long-term viability of the organizations. Proper definition of membership and leadership are mechanisms for ensuring this accountability. Identification of good leadership is also important. This has to be ensured while handing over management under PIM to WUAs (CWRDM, 1998).

3. **Legislative Backing:** The provisions made in the CADA act of 1986 do not meet the requirements for implementing PIM in Kerala. Even though there exists provisions for formation of farmer associations, canal and project level committees under the act, no specific powers have been given for carrying out duties assigned to them. Also, there is no provision in the act for agreement regarding supply of water to farmer associations. The same is the case with regard to the Kerala Irrigation and Water Conservation Act 2003. The Chapter on PIM provided in the Act is incomplete. There is only one tier organization mentioned in the Act, namely, WUAs at the outlet level. Transfer of the irrigation system, agreement between government and WUAs, etc. are not mentioned in the Act. It is therefore essential to either amend the existing Irrigation and Water Conservation Act or enact a new PIM Act.
4. **Clear Water Rights:** Under the existing physical infrastructure of irrigation projects, water rights could be defined as volume of water at the main canal, share of flow at the head of tertiary units and time duration within the command area of the watercourse (Strosser, 1988). Water is treated as the property of the government and most of the irrigation acts in India proceed from the basic premise that water is not a right of farmers. However, under PIM, there has to be a clear concept regarding allocation of water to the WUAs (either as a defined volume or defined proportion), which they purchase for a fee, and distribute among farmers as per accepted rules of equity and sustainability. The water quota for an average year may be decided on the basis of provisions made in the concerned project, and as declared by the general policy guidelines issued by the State government. The document on amendment to irrigation act for PIM of the Government

of India (MOWR, 1998) directs that it is necessary for irrigation acts to permit long-term allocation of water to WUAs and incorporate such allocations in the agreements with them.

Under PIM, WUAs should be given freedom in deciding their cropping pattern, within the quantum of water allocated under the operational schedule of the irrigation canal transferred to them. At the same time, they should also have some powers to restrict the type of crops grown by its members, taking into consideration the crop water requirements and need for equity in water distribution among members.

5. **Rehabilitation of the irrigation system:** One of the major reasons for the low efficiency of irrigation projects is lack of proper maintenance of the physical system, primarily due to lack of sufficient government fund. Hence, proper rehabilitation of the system before irrigation management transfer to farmers is absolutely essential in order to ensure water availability, and accordingly, motivate farmers to implement PIM. It should also be ensured that rehabilitation involves farmers in its various stages, and that the work is compatible with farmers' needs.

6. **Meaningful transfer agreement:** The transfer of irrigation system to farmers under PIM has to be preceded by a number of activities, including preparation of the inventory of works, structures and equipment to be transferred, joint inspection of these by WUAs and irrigation agency, carrying out necessary testing and repairs at the cost of the irrigation agency, and finally, handing over the management of the system, along with all necessary documents to the WUAs. The transfer agreement should also include provisions stating that the system transferred is in good working condition.

   The transfer agreement through a memorandum of understanding (MOU) should specify the structures transferred by the agency to WUAs, the area of operation, the rights and responsibilities of both the parties, the legal support available, the conditions regarding remitting of water charges by the WUAs and the agreement from WUAs to carry out the responsibilities assigned to them.

7. **Co-ordination between stakeholders:** The main stakeholders of an irrigation project are farmers and officials. One of the important factors responsible for low performance of irrigated agriculture has been found to be the unwieldy and inefficient bureaucracy spread over various departments, operating within complex departmental procedures and maintaining very less interaction with farmers. Under CADA set up, activities are restricted to below the outlets. Hence, the officials are not in a position to influence water availability, which falls under the purview of the irrigation/water resources department. Hence, while implementing PIM, the concerned officials should be in apposition to co-ordinate their activities, in line with the requirements of farmers, leading to the delivery of adequate and timely water to the WUAs.

Similarly, officials should develop a positive attitude towards better interaction/communication with farmers, and also ensure that this is achieved through conducting suitable training/awareness programmes.

8. **Consultation and involvement of farmers:** Inculcating a sense of ownership is an essential ingredient for effective people's participation. This requires sufficient consultation with farmers and enlisting their involvement in different stages of PIM such as organizing farmers, planning, operation and maintenance, monitoring etc. There are examples of successful farmer participation under minor irrigation in the Kerala Community Irrigation Project (KCIP) and the Kerala Minor Irrigation Project (KMIP) because there were provisions for prior consultation and involvement of farmers under these projects.
9. **Motivation and incentive systems:** Package of incentives for farmers as well as officials to make the management transfer programme effective can be expected to yield good results under PIM. Sufficient encouragement should be given to officials, who do good work though awards/ honours. It is also required to eliminate apprehensions of officials regarding aspects such as job insecurity, professional status, new responsibilities etc. when PIM is implemented. Similarly, farmers should also be motivated to perform their duties properly through provision of relevant incentives. Further, disincentives for free rider farmers should also be incorporated under the programme.
10. **Training/awareness programmes:** It is required to bring in an attitude change among officials, orienting them towards a set up, wherein, farmers will play a major role in managing irrigation projects. They should also be properly exposed to aspects such as knowledge and skills related to improved irrigation management, PIM related procedural, administrative and organizational issues, available sources of incentives and funding, commercial matters and accounting procedures. Similarly, managerial and technical skills have to be developed among farmers and farmer leaders to take on their new roles under PIM. This requires training and awareness programmes, workshops involving farmers and officials, and exposure visits to projects where PIM has been successfully implemented.
11. **Involvement of women:** Most often men are found to manage irrigation. Women's participation in this line has not been given due attention, even though they often provide considerable assistance in agriculture-related activities of farming households. Throughout most of the developing world, women lack either ownership or effective control of land, water and other natural resources, even though their labour is widely used. Women, actually play a very important role in irrigated farming, but it is mostly indirect and invisible. Many of the problems in irrigation management can be tackled if women are equipped properly. Studies (CWRDM & CADA, 2001) have found that women feel they can involve in various aspects of canal management

such as maintenance of the system, water distribution and settlement of disputes. Male farmers would also take into consideration many of the suggestions offered by their wives. Hence, emphasis on involvement of women would add to the sustainability of WUAs under PIM. A gender assessment study conducted in Thrissur district of Kerala under the KCIP, the first irrigation project in Kerala to involve women farmers, has revealed that when only 25% of men depend on agriculture; about 46% women are involved in it. It has also been found that more than 75% of the men farmers are part-time agriculturists, who depend on other vocations for their livelihood, and hence, have limited interest in farming (Chackacherry, 1996). In the training programme organized under KCIP, the main participants have been women. They were also found to show interest in knowing about improved agronomic practices. Further, about 45% of members of WUAs under KCIP were women. Chackacherry (1996) reported that men do not have any problem in bringing women into irrigated agriculture. In short, efforts to orient women and equip them to take on responsibilities and tasks under PIM can be expected to offer considerable help in implementing irrigation management transfer in Kerala.

12. **Mechanisms to settle disputes:** Many of the problems in irrigation management arise due to lack of transparency in the functioning of the irrigation system as such. Of prime importance in this regard is the awareness on water account. Aspects such as the quantity of water available in the reservoir / main system, how much will be available to the users, the policy directives of government which influence water availability/use etc. are to be made aware to the farmers by the PIM implementing agency. Designated officers of the agency should be given powers to penalize offending irrigators. Most importantly, the WUA should be empowered to stop supply of water to a user, in case an offence is committed by him/her. PIM will involve some form of sharing arrangement for water supplies between members. This can be helpful in building up good social capital of trust and co-operation between farmers, a condition naturally favourable for settlement of disputes among them.

PILOT PROJECTS ON PIM

As mentioned earlier, Kerala lags behind many states in India in participating farmers in the management of irrigation, and implementing PIM. Based on the outcome of the study (CWRDM & CADA, 2001), two pilot projects on PIM were formulated. They are now being implemented in Neyyar (mostly catering garden land crops) and Malampuzha irrigation projects in the State (mostly catering rice), with the objective of evolving a strategy for implementation of PIM in all the irrigation projects of Kerala, by demonstrating its possibility and convincing officials, farmers and others concerned. The programme envisaged is to transfer the operation and maintenance of one branch canal each at Neyyar Irrigation Project

(Olathanni branch canal–6.41 km length, 501 ha of ayacut area with garden land crops) and at Malampuzha Irrigation Project (Kuthannur branch canal–14.63 km length, 1664 ha of ayacut area with rice crop) to the farming community. The strategy and methodology adopted, and the phasing of the pilot projects are based on the suggestions of the study referred above. Since the existing farmer organization structure and government set up are not congenial for the implementation of PIM, changes are brought in on these aspects. A three-tier system with outlet based Water User Associations (WUAs), branch/distributory level WUAs, and project level Project Management Council is the structure tried in the farmer organization set up. Overseer, Assistant Engineer/Assistant Executive Engineer, and Executive Engineer respectively, will be attached as competent authority to these associations to help them in implementing their decisions. Works in the irrigation systems will be identified, prioritized and implemented by the WUAs. The competent authority will render technical advice and ensure that the works are carried out as per technical specifications. During turnover of operation and maintenance to WUAs, the WUA concerned and Water Resources Department will enter into a MOU, which clearly mentions about the commitments of both parties. Encouraging group farming, bringing women to the mainstream of irrigated agriculture, establishing relations with panchayaths, creating opportunities for coordination among the departments/agencies concerned, mobilizing tie-ups with marketing establishments, etc. are other allied activities carried out under the project. The programmes are carried out through five different phases, namely, Preparation, Organisation, Rehabilitation, Capacity Building and Turnover phases. Presently the pilot projects are in the last part of the Rehabilitation phase.

CONCLUSION

Till recently, irrigation management was considered as a technical or engineering subject, but it is now looked upon more and more as an integrated socio-technical enterprise, involving not only technical skills but also managerial tasks relating to the irrigation system. Irrigation management in almost all countries has been undergoing a great change during the last two decades. Management practices of the past era are no longer sufficiently cost effective, efficient in water use, sustainable, environmentally sound, or in line with trends for decentralization, privatization and accountability to users. Countries that once promoted more and more government involvement in irrigation management are adopting new policies that do just the opposite. Therefore, PIM is considered as a strategic intervention to improve the irrigation sector and mitigate the existing problem of farmers, water managers, and government. Interestingly, PIM is seen as a means to relieve budgetary pressures, especially when the cost of deferred maintenance cumulates year after year. Apart from the prerequisites mentioned in this article, and the strategy worked out from the study, what is more important is the fact that implementation of PIM needs to be carried out in a carefully staggered manner, which allows adequate time and effort in training and motivation, making use of the farmers' organizational skills and experience.

ACKNOWLEDGEMENT

We are extremely grateful to Dr. E.J. James, Executive Director, Centre for Water Resources Development and Management, Kozhikode, Kerala for all the encouragement given for conducting the study and preparation of this research paper.

REFERENCES

Central Water Commission (1992), *An Approach to Organisational and Procedural Changes in Irrigation Sector*, Report of Central Water Commission, New Delhi.

Centre for Water Resources Development and Management (1998), *Evaluation of Community Irrigation Systems*, Final Report submitted to the State Committee on Science, Technology and Environment, Government of Kerala.

Centre for Water Resources Development and Management (1999), *Evaluation of Beneficiary Farmers' Associations under Command Area Development Programme in Kerala*. Final Report submitted to INCID, New Delhi.

Centre for Water Resources Development and Management and Command Area Development in Kerala (2001), *Implementation of Participatory Irrigation Management in* Kerala, Final Report of the Study on Re-organisation of Farmer Associations for Implementing PIM in Kerala, submitted to Government of Kerala.

Chackacherry, George (1995), *Trends in Decentralisation in Water Management in Kerala*, In Proceedings of the International Conference on Water Management, Chennai, India: Confederation of Indian Industry.

Chackacherry, George (1996), *Kitchen to Farm: Suggestions for an Action Oriented Plan and Working Paper on Women in Agro-Economic Activities*, Kerala Community Irrigation Project, Kerala.

Command Area Development Authority (2004), *Annual Report 2003 – 2004*, Kerala.

Food and Agriculture Organisation (1995), *Irrigation Management Transfer in Asia*, Papers from the Expert Consultation on Irrigation Management Transfer in Asia, RAP Publication.

Madhava Chandran, K. and Chackacherry, George (2004), Factors Influencing Farmer Participation in Irrigation Management, *Journal of Tropical Agriculture,* 42 (1-2) : 77–79.

Ministry of Water Resources (1996), *Report of the Working Group on Participatory Irrigation, Management for Formulation of Plan Proposals for Ninth Five Year Plan* (Draft), New Delhi.

Ministry of Water Resources (1998), *Amendment to Irrigation Act for Participatory Irrigation Management*, New Delhi.

Strosser, Pierre (1998), Water Markets in Pakistan, *INPIM Newsletter*, No. 8.

World Bank (1998), *Handbook on Participatory Irrigation Management*, Economic Development Institute, Washington.

27

SUSTAINABLE DEVELOPMENT OF SMALL WATER BODIES IN TAMIL NADU

K. Sivasubramaniyan

I

INTRODUCTION

The primary sector of agriculture in India contributes about 28 per cent of Gross National Product and around 65 per cent of employment. Nearly three fourths of India's population living in rural areas still depends upon this sector for their livelihood. Although India has achieved self-sufficiency in foodgrain production, by increasing it from merely 51 million tonnes in 1950-51 to about 209 million tonnes in 1999-2000, the level of productivity attained both in irrigated and unirrigated areas is much less compared to other developing countries. This is one of the major reasons for the poor living standards of Indian farmers. There are several factors responsible for the low productivity of agriculture in many states of India. One of the primary factors is the inadequate and unsustainable irrigation facilities created and managed by both the government and local communities. Even though, the irrigated area of the country has increased more than two-fold (from 22.3 million hectares to 53.5 million hectares) over the past five decades the rainfed areas still account for 62 per cent of the cultivated land and it contributes nearly 44 per cent to agricultural output.

Water is one of the crucial inputs to enhance agricultural production. Crop yields depend critically on the amount of water available to crops at different stages of their growth. Rainfall alone is an inadequate and undependable source of moisture to sustain high productivity of agriculture. In India, a greater part of the year is practically dry; and the available rainfall even during the monsoon is often inadequate and highly variable in its distribution. While some regions have excess rainfall often resulting in floods, others suffer drought. These are among the principal reasons for the low level of yield in Indian agriculture. Under these conditions irrigation plays a crucial role in supplementing rainfall. It augments the supply of water to crops, makes it more assured, extends supply into the dry season, helps raise crop yields, adopt more productive crop patterns and increase the intensity of cultivation. Good quality of irrigation in terms of adequate, assured and controlled supply of water according to crop needs is especially important to realise the full potential of new technology based on fertiliser intense high yielding varieties.

The expansion and improvement of irrigation facilities has been an important element of India's agricultural development programme. From the beginning of the Plan period up to the end of the VIII Plan (1951-97) the public sector has spent more than Rs. 790 billion on irrigation development (including 70 billion of institutional finance almost exclusively to support groundwater exploitation by the farmers) mostly on major and medium reservoir based projects (Vaidyanathan: 2001: 2). However, public investments in minor surface irrigation works (tanks and small diversion works serving relatively small areas over a few villages utmost) have been limited (which attracted merely 153 billion during the same period). The exploitation of groundwater has grown very rapidly: The number of wells and tube wells has increased manifold. At the time of Independence most wells were operated manually or with animal power. But now most of them are operating with electric or diesel pumps. The proportion of irrigated area served by wells has risen from 29.8 per cent in the 1950s to 52.3 per cent in the 1990s. The same trend has also been noticed across states of India and in particular in Tamil Nadu the contribution of wells has increased from 23.5 per cent to 44.9 per cent in the same period. Except for public tube wells in north India this expansion has been mostly in the private sector with liberal financial assistance (mainly loans at relatively low interest) from the public sector (see Vaidyanathan 1994a).

The quality of irrigation that is, the ability to regulate timing and quantity of water supply to particular plots according to crop needs varies between different type of projects. Large reservoirs, which harness rainfall over extensive areas, provide a larger volume, and more assured supply, than minor irrigation works like tanks which depend mostly on local rainfall. By contrast, in the case of wells which usually serve small areas, often belonging to a single farmer, it is easier to regulate water supply according to specific circumstances and needs of individual farmers. So, the quality of irrigation is better in the case of wells. However, wells are not always reliable in all areas, especially, where there is no surface source of supply available. In those areas well water is mostly depend upon by the availability of rainfall and this situation prevails in most parts of the country. In order to overcome this difficulty the rain harvesting method was encouraged by the government. This method is useful to store water which is available in a smaller catchment area through construction of tanks. But these tanks are not altogether feasible to store water which is available from the larger catchment through big river basins and perennial streams. In order to utilise the river water the governments took initiative to build dams and reservoirs to store available water which was mainly used for irrigation other than Power generation. Hence the growing role of reservoir based surface irrigation and wells have helped to improve the quality of irrigation (Sivasubramaniyan 1995).

A sound understanding of the pattern of agricultural development through expansion of different sources of irrigation is important mainly to provide preference over one source to other which would help to develop further the preferencial source in a complementary manner. In short, among different sources (canals, tanks, wells and other sources) of irrigation, tanks (otherwise called small water bodies) are considered as the prime source for the development of agriculture. Because this source also indirectly helps the wells to get recharge its supply. Even then, due importance has not been given over the Plan periods to develop this source in a sustainable manner. As a result, the country has been facing water scarcity problems in most regions and also depriving to recharge wells adequately. There are number of studies available on tank irrigation but most of them are preoccupied to focus its attention in one or two aspects only. Hence, in this paper, an attempt has been made to provide a comprehensive view of smaller water bodies such as tanks, their present condition, necessity and importance to be given to develop this source all over India to perform our agriculture in a sustainable manner. The main focus of the paper is to: (i) provide a comprehensive view of tank irrigation and its management based on available studies; (ii) find out the reasons for the decline of tank irrigation in Tamil Nadu and (iii) suggest ways to revive this source to its previous position, in terms of the area irrigated by this source, say a million hectares in Tamil Nadu. The paper is organized in the following fashion. Section II reviews the relevant literature on irrigation institutions (the base for sustainable development of tanks); Section III discusses user participation in management of irrigation systems in India and other parts of the world; Section IV provides a comparative picture of different sources of irrigation in India as well as in Tamil Nadu in a historical perspective and importance of tanks in the three southern states of India; Section V examines the major problems faced by the tank irrigation system; Section VI is devoted for policy recommendations for improvement and restoration of tanks and finally Section VII outlines the conclusions.

II

REVIEW OF LITERATURE

The quality of irrigation is not merely a function of the nature of the water source but of how well the source is managed. Partly this is a matter of engineering: For example, the ability to regulate the amount of water supplied to different segments of the command depends on how well the canals and distributaries are designed and constructed, and whether they are equipped with regulatory devices. But to a large extent these matters also depend on having a clearly defined and consistent set of rules to guide the allocation between various uses and users, and the ability to enforce these rules. All surface systems typically serve numerous users ranging from a few dozen farmers in the case of small tanks to hundreds of thousands in projects like Bhakra Nangal. In these cases, the scheduling of supplies between time and space is especially complex. This is where institutions come in. Apparently, the reservoir based irrigation attracts a large number of communities to be involved to utilise the available water. In order to regulate and manage the surface irrigation systems (whether it is small or big) separate irrigation institutions (IIs) have to build up.

IMPORTANCE OF IRRIGATION INSTITUTIONS

The early work on IIs was done by Wittfogel (1957), who concerned with the relation of irrigation with the emergence of strong centralised state systems. He argued that hydraulic agriculture, which requires great cooperative effort for organising capital and labour to build dams and canals, leads to greater political integration and centralisation. Following Wittfogel there have been several studies which have questioned the universal applicability of his irrigation hypothesis (large scale irrigation system requires great cooperative efforts to build dams and canals which in turn leads to greater political integration and centralisation). Thus, Leach (1961), Millon (1962), and Maass and Anderson (1978) found no clear cut relationship between the degree of centralisation and the size of irrigation system; nor any evidence that irrigation works invariably result in a hydraulic bureaucracy or strengthening of the centralised power. However, there is some confusion in the concept of centralisation. Hunt and Hunt (1976: 129) have argued the need to distinguish between two distinct types of centralisation: "One refers to authority in terms of irrigation systems, who exercised different decisions of the social and technical needs of irrigation. The other refers to generalised political authority, who exercised water as one aspect of a complex political role, which may involve other functions beyond water control." Whether and to what extent irrigation development contributes to centralisation of the state is evidently a debatable issue. But our concern here is with the way irrigation operations are structured and function and how they affect agriculture.

As it is indicated at the outset water is the basic input to raise crop productivity. Here it is important to point out the peculiarities of water and its use in agriculture. Water is in the nature of common resource serving numerous uses and users. Harnessing and using this resource for common benefit of users is generally done (or regulated) by some form of public organisation. Private ownership and operation is rare except in the case of wells which serve small areas and are within the means of individual farmers. In surface irrigation the arrangements range from exclusive user management of construction maintenance and water allocation to exclusively government managed systems, with a wide variety of mixtures in between.

In principle, since irrigation contributes to a significant increase in productivity, the potential beneficiaries have collective interest to develop and manage it. The world over, there are innumerable examples of such efforts in developing small works (like tanks, diversion of streams and rivers) which involve relatively simple engineering and serve relatively small, compact community of users. But in larger systems involving complex engineering, large outlays and vast number of users, the state tends to play a greater role in financing, constructing and managing them. The organisations in such cases are necessarily larger, more complex, more dependent on formal rules and bureaucracies. Indeed in several countries, including India the state has taken over the full responsibility for all these activities in the case of modern reservoir based irrigation. Even in the case of groundwater, which is exploited privately, regulation to prevent exploitation requires public (State) intervention.

The relative roles of the government, the user community and individual farmer, and the inter relation among them obviously vary a great deal. This is reflected in the form of the organisation, its internal structure (including the distribution of responsibilities and powers within the organisation; the locus and modalities of decision making; the manner in which the functionaries are appointed and controlled) and the relation between irrigation organisations and the wider socio political framework. However, it is inappropriate, and even misleading, to view irrigation organisation as if it were a single entity: rather it has to be viewed in terms of arrangements for performing various tasks such as facilities construction, operation maintenance, water allocation and conflict resolution in different phases of irrigation management (Kelly 1980: 14 20).

These variations are a function of several factors: Rainfall, geology, topography and other features of the physical environment have a significant bearing on the kind of systems (both in terms of water source and scale) which are desirable and feasible in a particular context. The scale and complexity of the system affects whether or not it can be handled on an individual basis or by cooperative effort on the part of users or needs State involvement. The extent of State, users in terms of their sharing in the costs of setting up and operating systems have influence their interest and involvement. Often the interests of politicians and the bureaucracy dictate a high degree of direct role for the state in funding and managing irrigation[1].

Comprehensive descriptions of the structure and functioning of particular systems in a wider, historical perspective are rare; systematic comparative studies are even more so[2]. The literature on IIs has focused more on issues as centralisation vs decentralisation, bureaucracy vs user management, privatisation, and cost recovery. This preoccupation has been strengthened by the growing concern about the poor quality of service, depletion of groundwater; salinity, water logging and environmental problems; low cost recovery and similar features of State managed systems, which have become increasingly important in surface irrigation.

However, it is not, as if users have been indifferent to the way systems are managed. Studies have shown the existence of collective effort among users of even State run canal systems. Wade (1987a) found that scarcity of water, cooperation among farmers and specialised irrigation were characteristic of the Tungabadhra canal system of Andhra Pradesh. Further he stated that there are strong and active village committees dealing with irrigation problems. These committees collect funds, appoint common irrigators and sluice guards to patrol distributaries at the top end of canals, arrange representations to officials and politicians and has a sizeable common fund for different expenses. Many other cases of local mobilisation to persuade and/or pressure the state to construct/improve irrigation works, to change operational rules, ensure adequate supplies from the main system especially in times of scarcity have been reported (Vaidyanathan and Janakarajan (1989); Rajagopal 1991; and Sivasubramaniyan 2000a). But, instances of the users actively participating in the design, framing of operational rules or continuing management of the system as a whole are few and far between.

III

USER PARTICIPATION IN MANAGEMENT

There is good reason to believe that centralised bureaucratic management with little or no user participation leads to inefficient irrigation, which contributes to the poor quality of irrigation provided by state run systems. This has long been emphasised in irrigation management literature (see Coward 1980; Downing and Gibson 1974). Recent studies underscore this point even more forcefully (see Vaidyanathan 1991; Chambers 1988; Sengupta 1991; Ostrom 1992; Tang 1992; GOI 1992; IIMI 1994, Sivasubramaniyan 1985).

A recent work (Tang 1992), based on analysis of 47 case studies of irrigation systems around the world, highlights the wide diversity of experiences in the relationship between institutions and collective action in resource management. The results suggest that systems that include elements of self governance (by users) can work well. As Tang points out, how well each system works depends on the physical, social and institutional factors operating in each case. These factors interact to create the unique set of needs and problems that must be handled by each governing institution. On the basis of a comparative study of India and Philippines Sengupta (1991) has argued that farmers participation is feasible in practical terms, and beneficial whenever applied. In a subsequent study (Sengupta 1993) based on data for over fifty such systems in India, and their aggregate effects, environmental impact, multipurpose uses and salient management features, he has suggested that a suitable alternative strategy can be developed based on the existing traditional techniques of irrigation and water management. He has also suggested that such an approach would promote efficient water management, better flood control and soil conservation, farmers' participation and balanced agricultural development.

In India, the need for wider involvement and participation of users for effective management has been long recognised. Encouraging users involvement is now part of official policy. The necessary enabling legislation has also been enacted in several States. A number of experiments in this direction have also been made under the auspices of non governmental organisations in different parts of the country. However, a recent assessment of this experience has concluded thus: "Efforts to actually organise farmers' groups and make them participate in management have been patchy and can claim only a limited success so far. Most State Governments have not really pursued the idea seriously" (GOI 1992: 125). However, Maharashtra and Gujarat have done considerable pioneering work, both through non governmental and by the government organisations. The Planning Commission evaluated on this aspect thus: Overall "... the area covered by these initiatives is very small, less than one per cent of area irrigated at present... For the most part, the outlet and canal committees are there only in name; their functions are vague; they seldom meet; they are not consulted on substantial issues; nor are department officers required to follow the advice (GOI 1992: 126).

Elsewhere in the world, similar experiments[3] have been made or are under way. They vary in scope some have set up local water users associations; some have tried to involve them in the management of the system as a whole; and in a few cases more radical reforms seeking to get users to participate in system design, contribute to the costs and take responsibility for maintenance and operation, are being attempted.

The conference strengthened the growing consensus to move towards greater involvement of users and to substantially reduce the role of government in constructing and managing irrigation systems. But such reform not only requires the creation of Water Users Associations but also to strengthen the Associations at the tertiary level. Major role for users in decision making and greater responsibility for carrying out various management *tasks* such as construction of irrigation structures, operation and maintenance, water allocation and resolution of conflicts in all *phases* of irrigation system such as water source control, water delivery, water use and water drainage is necessary. As Ostrom (1992) has pointed out that involving users and suppliers directly in all stages (including design) helps ensure that institutions are well matched to the particular physical, economic and cultural environment of each system. Such "self governing" systems are more flexible, distribute water more equitably, and neutralize the "perverse incentives" that lead managers and consumers to abuse and neglect irrigation facilities.

The question of moving towards self governing form of irrigation organisation with user involvement is important in the case of relatively large systems now under State control and even smaller, multi community systems where the State plays an important role. But the problem is rather different in the case of local surface irrigation works which have a tradition of community management. These works are very numerous and account for a sizeable proportion of the irrigated area in India. While their problems are different, their experience in user management the manner in which they are managed, the way they have changed over time, and their problems in ensuring efficient and equitable water management are of considerable practical interest to institutional reform in larger irrigation systems.

MANAGEMENT OF LOCAL SYSTEMS

India has a long tradition of using small storage and diversion canals drawn from local streams and rivers. Before the introduction of large reservoir based canal irrigation and energised pumping of groundwater, tanks were the dominant form of irrigation in several parts of the country. In the 1950s tanks and other minor sources irrigated 6.5 million hectares, which constituted about 29 per cent of the country's total irrigated land. Since then the bulk of the extension of irrigation has been through large reservoirs and groundwater. The relative importance of minor irrigation, according to official statistics, even the absolute area under this category has declined. This is attributed partly to the concentration of investment on large storages and groundwater; and partly to the deterioration of traditional local systems (Vaidyanathan 1992: 2 5).

IMPORTANCE OF MINOR IRRIGATION TANKS

Rainfall is the only source that brings water to the earth, but nature's gift is not always utilised properly. In India, with the available annual precipitation of 4000 KM^3, about 51 per cent is not controllable due to evaporation and 49 per cent goes as runoff. From the latter, currently 35 per cent of surface flow and 20 per cent of ground water are estimated as utilizable water resources. But actually, at present only 30 per cent (600 KM^3) of both the surface and groundwater resources are brought under utilization and the remaining 70 per cent need to be developed.

The total capacity of large storages in India at the time of Independence was 18 KM^3 (636 TMC). It increased to 250 KM^3 in 2000. The total capacity of tanks and ponds in the 1970s was 15 KM^3 or 530 TMC. But its capacity has not increased much after that period till now. However, there is wide scope for increasing its capacity if we devote a minimum level of 3 per cent of total area which fetches an increase in capacity of 400 KM^3. Actually this capacity is much more than the total capacity on all large storages in our country. And this increase is about 26 times more than the present capacity of all tanks in India.

TANKS IRRIGATION IN SOUTH INDIA

Among minor irrigation works tanks are the most important category. The minor irrigation census conducted in 1985-86 estimated that the country has some 2,50,000 tanks (including tanks-not-in-use). The reported area under tank irrigation was about 3.5 million hectares in the mid eighties as against 4.5 million hectares recorded in 1960 61. Tank irrigation is particularly widespread in South India. Nearly 60 per cent of the area under tanks is in Tamil Nadu, Andhra Pradesh and Karnataka. In these three states, tanks irrigate a much higher proportion of the total cultivated land (seven per cent against the national average of two per cent) and of the total irrigated land (24 per cent as against the national average of eight per cent) (GOI 1989).

Most tanks in south India are of ancient origin, with several dating back to the 4th and 5th centuries AD. Tamil Nadu, Andhra Pradesh and Karnataka have nearly 1,20,000 tanks (out of the 2,08,000 tanks-in-use in the country as a whole) irrigating 1.8 million hectares. During 1997-98 to 1999-2000, the area irrigated by tanks in these three states taken together comprised a little over sixth of the total irrigated area and 6 per cent of the total cultivated area. By both indicators tanks are more prominent in Tamil Nadu and Andhra Pradesh than in Karnataka. The distribution of tank irrigation, as well as its importance as a source of irrigation within the region, is highly variable. Nearly 70 per cent of the total *ayacuts* of tanks in the three states is concentrated in 19 districts (out of 60 districts); two-fifths is accounted for eight districts (Vaidyanathan and Sivasubramaniyan: 2001).

TANKS IN TAMIL NADU

In Tamil Nadu, where somewhat more detailed information on size distribution of tanks is available, nearly two-fifths have *ayacut* of less than 20 ha. more than half serve less than 40 ha. There are only 534 tanks (less than 2 per cent of the total number) with *ayacut* exceeding 200 ha; but these tanks are estimated to irrigate 2,56,000 ha. or nearly 25 per cent of the total area (Table 1). There are a few large tanks with *ayacuts* exceeding 1000 hectares, the largest of them serving 6,000 ha. (Vaidyanathan and Sivasubramaniyan 2001). Official figures indicate, Tamil Nadu has some 39,200 tanks of which about half are small (serving less than 40 ha. each) and depend solely on local rainfall. There are about 5300 rainfed tanks with command exceeding 40 ha. Some 3700 tanks are linked to larger systems fed by canals drawn from large rivers and storages (See Table 2). These data should, however, be interpreted with much caution. The PWD has no systematic arrangements for maintaining an up-to-date inventory of tanks and their *ayacuts*. Available compilations tend to be outdated. Moreover, the number of tanks as reported in the Season and Crop Reports vary over time; and the net area irrigated by them vary and in general are much below the PWD figures of registered *ayacut*.

Though the tanks are generally small, each serving one village or a part of it; some are large spread over several villages serving several thousand hectares. However, the volume and duration of water supply relative to area irrigated, as well as its assurance of water supply vary widely among them. We have some sketchy accounts of their historical origin and evolution (see Raju 1941; Ludden 1978; Reddy 1990). For the

Table 1
Districtwise Size-distribution of Tanks in Tamil Nadu, 1977

Sl. No.	*District*	*Panchayat Union Tanks*						*P.W.D. Tanks*					
		<20 ha		*20-40 ha*		*Total*		*40-200 ha*		*200-400 ha*		*400-800 ha*	
		No.	*Ayacut*	*No.*	*Ayacut*	*No.*	*Ayacut*	*No.*	*Ayacut*	*No.*	*Ayacut*	*No.*	*Ayacut*
1	2	3	4	5	6	7	8	9	10	11	12	13	14
1.	Chengai-MGR	1241	4858	542	13158	1783	18016	1088	95071	114	30726	29	15140
2.	North Arcot Ambedkar + T.V.Malai Sambuvarayar	1482	5668	602	14615	2084	20283	1116	74457	46	12183	5	3030
3.	South Arcot	1213	4858	553	13441	1766	18300	692	43575	47	13160	10	4733
4.	Salem	449	1619	100	2429	549	4049	181	14855	7	1677	0	0
5.	Dharmapuri	1451	5668	128	3117	1579	8785	95	6970	5	1257	0	0
6.	Coimbatore + Periyar	42	162	22	526	64	688	52	5252	6	1743	1	577
7.	Thanjavur	338	1215	153	3725	491	4939	663	22532	13	3753	6	3030
8.	Tiruchirapalli							238	16057	12	3288	3	1522
		4609	18462	725	17611	5334	36073						
9.	Pudukkottai							508	32300	21	5197	1	668
10.	Madurai + Anna	3142	12551	249	6032	3391	18583	731	41855	32	9222	6	3322
11.	Ramanathapuram + PMT + Kamarajar	642	2429	691	16761	1333	19190	1419	97360	56	17244	24	12084
12.	T.Veli Kattabomman + Chidambaranar	806	3239	159	3846	965	7085	642	35430	26	4809	8	4412
13.	Kanyakumari	1062	4049	12	283	1074	4332	980	9435	4	1168	0	0
14.	Nilgiris	0	0	0	0	0	0	0	0	0	0	0	0
	Total	16477	64777	3936	95547	20413	160324	8405	495147	389	105426	93	48520

Contd...

Table 1 Continued.... **(Ayacut in hectares)**

Sl.	District	800-2000 ha		>2000 ha		Total		Ex-Zamin Tanks		All Tanks Total		Ayacut per tank
		No.	Ayacut	No.	Ayacut	No.	Ayacut	No.	Ayacut	No.	Ayacut	
1	2	15	16	17	18	19	20	21	22	23	24	25
1.	Chengai-MGR	9	9082	3	9931	1243	159950	720	7328	3746	185294	49
2.	North Arcot Ambedkar + T.V.Malai Sambuvarayar	1	1627	1	2315	1169	93612	482	4899	3735	118794	32
3.	South Arcot	4	4939	4	35310	757	101716	79	810	2602	120826	46
4.	Salem	0	0	0	0	188	16532	0	0	737	20581	28
5.	Dharmapuri	1	970	0	0	101	9197	154	1579	1834	19562	11
6.	Coimbatore + Periyar	0	0	0	0	59	7571	0	0	123	8260	67
7.	Thanjavur	2	2240	1	2150	685	33704	0	0	1176	38644	33
8.	Tiruchirapalli	3	3406	2	4133	258	28407	58	607	6394	105438	16
9.	Pudukkottai	0	0	0	0	530	38165	214	2186			
10.	Madurai + Anna	2	2136	0	0	771	56535	331	3360	4493	78478	17
11.	Ramanathapuram + PMT + Kamarajar	8	9578	1	4916	1508	141181	7367	75101	10208	235473	23
12.	T.Veli Kattabomman + Chidambaranar	10	9309	0	0	686	53960	445	4534	2096	65579	31
13.	Kanyakumari	0	0	0	0	984	10602	0	0	2058	14934	7
14.	Nilgiris	0	0	0	0	0	0	0	0	0	0	0
	Total	40	43286	12	58754	8939	751133	9850	100405	39202	1011862	26

Table 2
Systemwise Details of PWD Tanks in Tamil Nadu, 1977

Sl. No.	*Name of the Basin*	*Name of the System*	*Districts benefited*	*No. of tanks*	*Ayacut (Ha)*	*% to total Tanks*	*% to total Ayacut*
1.	Cauvery	Lower Coleroon Anicut System	South Arcot	3	20224		
	Cauvery	Cauvery Delta System	Thanjavur	25	6139		
	Cauvery	Cauvery Mettur Project	Thanjavur	583	24746		
			Pudukkottai	161	8206		
	Cauvery	Kattalai Scheme	Thanjavur	8	725		
			Tiruchirapalli	11	1778		
	Cauvery	New Kattalai High Level Canal	Thanjavur	64	1643		
			Tiruchirapalli	44	2457		
	Cauvery	Pullambadi Canal	Tiruchirapalli	28	6213		
	Cauvery	Nandiyar Channel System	Tiruchirapalli	2	2041		
	Cauvery	Kodiveri Anicut System	Periyar	2	189		
	Basin total			931	74362	25	34
2.	Palar	Palar Anicut System	Chengai-MGR	71	9406		
			North Arcot Ambedkar	241	22528		
	Palar	Ponnai Anicut System	North Arcot Ambedkar	128	9120		
	Palar	Cheyyar Anicut System	Chengai-MGR	3	180		
			North Arcot Ambedkar	140	7883		
	Basin total			583	49117	16	23
3.	Periyar-Vaigai	Periyar System	Madurai	385	23999		
	Periyar-Vaigai	Vaigai Reservoir System	Madurai	98	2664		
			PMThevar	130	2598		
	Basin total			613	29262	17	14
4.	Tambraparani	Marudur Anicut System	Tirunelveli Kattabomman	31	5264		
	Tambraparani	Srivaikuntam Anicut System	Tirunelveli Kattabomman	23	7348		

Contd....

Sl. No.	Name of the Basin	Name of the System	Districts benefited	No. of tanks	Ayacut (Ha)	% to total Tanks	% to total Ayacut
	Tambraparani	Manimuthar Reservoir System	Tirunelveli Kattabomman	343	9319		
	Basin total			397	21932	11	10
5.	Vellar	Mehamathur Anicut System	South Arcot	11	1907		
	Vellar	Tholudur Anicut System	South Arcot	25	525		
	Vellar	Pelandurai Anicut System	South Arcot	18	3150		
	Vellar	Vridhachalam Anicut System	South Arcot	13	2214		
	Vellar	Sethiathope Anicut System	South Arcot	8	10603		
	Basin total			75	18399	2	9
6.	Ponnaiyar	Sathanur Reservoir System	North Arcot Ambedkar	28	649		
			South Arcot	10	399		
	Ponnaiyar	Thirukoilur Anicut System	South Arcot	96	7903		
	Ponnaiyar	Barur Tank System	Dharmapurai	1	970		
	Ponnaiyar	Krishnagiri Reservoir System	Dharmapurai	2	264		
	Basin total			137	10186	4	5
7.	Kodayar	Kodayar System	Kanyakumari	960	7768	26	4
8.	Madras (Coovam)	Chembarambakkam (Tank System)	Chengai-MGR	1	5271		2
		Total		3697	216296	100	100

Source: GOTN, Chief Engineer for Minor Irrigation, PWD, Madras, 1977.

most part tanks were constructed long ago (in some cases centuries ago) by efforts of local chieftains, temples and village communities. These tanks were maintained and managed by user communities. The *Kudimaramath* (a system of voluntary labour participation to upkeep and maintenance of irrigation channels) system provided for collective maintenance by community efforts: Under the *Mirasi* system of land tenure, a part of the village produce was ear marked for tank maintenance.

The wars led by local chieftains preceding the advent of British rule caused considerable physical damage to the irrigation works. Changes in land tenure and in particular the introduction of the Ryotwari system in place of *Mirasi* tenure and the recognition of ex rulers/chieftains as *Zamindars* have further weakened the traditional socio economic structures which took care of the tanks. The deterioration in irrigation adversely affected government revenues land revenue was the dominant source and this led the colonial government to assume direct responsibility for maintenance of system facilities (other than distribution networks) and subsequently for repair, renovation and improvement, construction of new tanks (Raju 1941; Krishnaswami 1947; Dikshit et al. 1994).

The government take over of maintenance did not, by all accounts, make much of a difference, partly because of inadequate allocation of resources. Attempts to revive *Kudimaramath* were also unsuccessful. However, the government did undertake substantial investments during the late 19th and the early 20th centuries for major repairs, construction/ remodelling of anicuts to augment supplies to tanks, increasing the number of system tanks and construction of new tanks. Such works were taken up in Palar, Vaigai and Tambraparni basins.

In the early 1900s, the government had initiated a survey of existing tanks to ascertain whether they were up to required standards and if not taken up works needed to bring them to standard. This survey made relatively slow progress. There was some quickening of pace in the post Independence period but outlays have been rather small and results moderate. During the 1980s a large tank modernisation project was started. This programme focused mainly on engineering works to improve tank structures and distribution networks more or less entirely under the Public Works Department (PWD). Lately the programme has emphasised participation of users in the tank modernisation projects. Surprisingly, however, neither government reports on minor irrigation nor the modernisation schemes, recognised that tank communities generally have informal local institutions for maintenance and operation. Nor do they explore the role which they might play in improving management.

COMMUNITY MANAGEMENT OF TANKS

That local irrigation works have traditional community based institutions for their management is well known and documented by studies in different parts of the world[4]. There are also several studies relating to India[5]. These studies, many of which relate to Tamil Nadu, are rather limited in scope and coverage, but nevertheless useful.

There are few properly documented accounts of the way these user communities worked before the advent of the British rule. It is known that they were governed by customs and tradition; in some cases (notably North Arcot district of Tamil Nadu) these have been codified in writing and published (Water *Mamulnamas*) in the early 19th century (Sivasubramaniyan 2000b). Many accounts in the latter part of the 19th century attributed the collapse of the *Kudimaramath* to the decline of these traditions. Currently too there is a widespread belief that traditional arrangements for tank management have disintegrated and become ineffective. But recent studies cited above show that while local tank management institutions are not uniformly effective, and there have been considerable changes, there is no warrant for the belief that they have declined.

Most communities studied are found to have well defined rules for the contributions by *ayacutdars* for maintenance and managing water allocation, and mechanisms for enforcing the same. In many studies it is found that the community was to be actively involved in collective action for appropriation of water (by participating in inlet channel maintenance to get increased supply), in maintenance of distribution networks in the *ayacut* where water is available, and in regulating water use especially in times of shortage through functionaries specifically appointed for the purpose (see Sivasubramaniyan 1998a, 1998b).

The general pattern observed in Tamil Nadu (for details see Vaidyanathan and Janakarajan 1989; Rajagopal 1991; Sengupta 1991, 1993; Ramanathan unpublished, Sivasubramaniyan 1997) is that the village leaders (*Nattanmaikars*) drawn from the large landholders of the dominant, usually upper, castes take active part in tank management. In some villages, the task is entrusted to a specialist functionary (*Kavaimaniyam*) experienced and knowledgeable about tank matters. Decisions made by them are implemented by irrigation workers (*Kammukatties*) mostly from the Scheduled Castes by rotation. They are expected to monitor water flows to the tank, watch the bunds and other tank structures, open and close sluices and regulate water flow in the *ayacut* when required by the *Nattanmai / Kavaimaniyam*. The *Kammukatties* are paid by the *ayacutdars*, usually in kind, on per acre basis. In the Tambraparni basin, the tank level organisations are more formal and elaborate (Sengupta 1991, 1993; Ramanathan unpublished).

There are well recognised procedures for maintenance of the distribution networks, including the contributions expected from the *ayacutdars*, penalties for non participation and the organisation of work. The *ayacutdars* are expected to contribute labour in proportion to their holding; sometimes cash contributions are allowed. In some cases the *ayacutdars* participate in cleaning inlet channels, when the PWD fails to do so. In some villages detailed accounts of contributions, default and fines are maintained.

There are also conventions and rules for regulating the flow of water in the *ayacut*. Based on a survey of about a dozen tanks in North Arcot district, Chambers (1977) noted the existence of systematic though varying rules for allocating water in times of scarcity. One village distributed water to fixed acreages during scarcity period; in five villages the usual distribution system of the "top enders first" was practised by the organisation, while in three others a special provision for "tail enders first" in times of scarcity was practised. Chambers reported however that water distribution under tanks was usually both inequitable and inefficient in terms of productivity. Because of conveyance losses water is less productive to the tail end of a channel. Scarcity of water also mostly affects the tail end farmers. The traditional rules of allocation according to the conditions prevailing then may not be applicable to the present. New rules may therefore emerge or changes may take place in the structure and functioning of earlier tank irrigation organisations.

Similar variations are recorded by Vaidyanathan and Janakarajan (1989) in their study on Palar Anicut System. According to them, there are considerable variations in the working of tank organisations: They found that while almost all villages had *Nattanmaikars* and *Neerkatties*, only five had *Kavaimaniyam*. Eight of the 15 villages covered by the study reported regular maintenance of inlet channels, distributaries and field channels. Six reported that regular maintenance was not done in the last five years. The system of rationing by turn in terms of scarcity was in vogue only in seven villages. The main features of villages where the institution functioned effectively include concentration of land control in one or two upper castes; absence of factions; tank being the only source of irrigation; low level of technology, and low absentee ownership (Janakarajan 1993b). The development of well irrigation and the fact that the 1980s were marked by a series of droughts and low tank supply must also have affected their functioning at the time of the survey.

A comparative study of canal and tank irrigation Systems in Cauvery Basin in Tamil Nadu has shown stronger institutional set up for collective action to secure supply in the systems with relatively poor, uncertain water availability compared to the system with abundant supply; greater uncertainty of supplies combined with greater difference between irrigated and unirrigated yield has stimulated greater interest in collective action in the lesser water availability system. The study also reported that the role of institutions in water management to be significant in the lesser water availability system where the degree of scarcity and consequently the potential productivity of water is high, but it is relatively insignificant as the system is not affected much by water scarcity. Even within an irrigation system, irrigation organisation is much stronger in the tail reach as it is more affected by water scarcity than others. This indicates that the extent of involvement of irrigation organisation in water management is linked to the extent of benefits likely to accrue to them by averting crop losses (Rajagopal 1991: 324 8).

The aforesaid studies also show that the characteristics of village level organisation have undergone changes because of vast increase in the number of *ayacutdars*, shift in land control from a relatively small number of upper caste land owners to numerous middle and lower castes, the attitude changes of farmers, and the spread of well irrigation in the tank *ayacut* of several villages. As a result, the arrangements have got modified but have not broken down. There is still a strong tendency to invoke customs and tradition in managing irrigation supplies. The necessity to invoke customs arises due to the fact that one could ascertain from the available data that the area served by tanks has been gradually declining not only in the tank intensive states but also all over the country since the late 1970s. Hence, it is pertinent to analyse the contribution of tank irrigation for sustenance of agriculture over a period of time.

IV

IRRIGATED AGRICULTURE AND SOURCE OF IRRIGATION

There have been a number of factors responsible for the persistent decline in the importance of tank irrigation over a period of 50 years in India and particularly in Tamil Nadu (TN). Before independence, this tank system which is one of the major components of Minor Irrigation sources was the major source of irrigation in many of the Indian states. Even now, in some of the districts of TN and also in the southern states of India as mentioned earlier, the contribution of tanks to both the total Net Irrigated Area (NIA) and Net Sown Area (NSA) is appreciable. Both in absolute and relative terms NIA by tanks in India and TN increased up to the 1960s, after that it declined steeply. However in TN, its percentage share to all India level remained more or less constant compared to the 1950s (see Table 3).

In absolute terms canal irrigation in Tamil Nadu is stagnant over the past five decades but in relative terms it has been gradually and steadily declining. The all India data show that while the absolute area has gradually increased from 92 lakh hectares (ha) to 173 lakh ha. the percentage share of canal irrigation has been marginally decreasing and this decrease is comparatively more in the 1990s than the other decades. Only in the case of wells and tube wells category a very steep increase in NIA is observed both in TN and at all India level over the past fifty years. This increase has been remarkable at all India level both in absolute as well as relative terms while in TN it has merely doubled. Contribution to NIA by 'other sources' in TN is negligible even from the 1950s and they have also been showing a steady decline both in absolute and relative terms. But at all India level this source has been contributing a considerable proportion compared to tanks. This has to be viewed in terms of increase in absolute area from 23.2 lakh ha. to 33.1 lakh ha. over a period of fifty years. In other words, during the 1950s a little over half of the area under tanks was contributed by other sources to total NIA. However, in the 1990s, the contribution of other sources is little higher than the NIA by tanks.

Table 3
Trends in Net Area Irrigated by Sources from 1950-51 to 1999-2000

Area in lakh hectares)

Source	1950-51 to 1959-60		1960-61 to 1969-70		1970-71 to 1979-80		1950-51 to 1979-80		1980-81 to 1989-90		1990-91 to 1999-2000		1980-81 to 1999-2000	
	Area	% to NIA	Area	% to NIA	Area	% to NIA	Area	% to NIA	Area	% to NIA	Area	% to NIA	Area	% to NIA
INDIA														
Govt. & private canals	91.9	41.2	111.9	41.9	137.7	40.1	113.8	40.9	163.1	38.3	173.4	32.7	168.3	35.2
Tanks	41.5	18.6	44.5	16.6	38.1	11.1	41.4	14.9	29.9	7.0	31.1	5.9	30.5	6.4
Wells+tubewells	66.3	29.8	87.1	32.6	144.1	41.9	99.2	35.7	207.8	48.7	292.5	55.2	250.1	52.3
Other sources *	23.2	10.4	23.9	8.9	23.8	6.9	23.6	8.5	25.4	6.0	33.1	6.2	29.2	6.1
Total NIA	222.9	100.0	267.3	100.0	343.6	100.0	277.9	100.0	426.3	100.0	530.1	100.0	478.2	100.0
TANMIL NADU														
Govt. & private canals	8.0 (8.7)	37.6	8.8 (7.9)	35.6	8.9 (6.5)	33.2	8.6 (7.5)	35.3	8.2 (5.0)	33.0	8.3 (4.8)	29.3	8.3 (4.9)	31.0
Tanks	7.8 (18.7)	36.8	9.1 (20.5)	36.8	8.5 (22.3)	31.5	8.5 (20.5)	34.8	6.2 (20.6)	24.7	6.3 (20.4)	22.4	6.3 (20.5)	23.4
Wells+tubewells	5.0 (7.5)	23.5	6.5 (7.4)	26.0	9.2 (6.4)	34.1	6.9 (6.9)	28.3	10.4 (5.0)	41.6	13.5 (4.6)	47.7	12.0 (4.8)	44.9
Other sources *	0.5 (2.0)	2.2	0.4 (1.6)	1.6	0.4 (1.5)	1.3	0.4 (1.7)	1.6	0.2 (0.7)	0.8	0.2 (0.5)	0.6	0.2 (0.6)	0.7
Total NIA	21.2 (9.5)	100.0	24.8 (9.3)	100.0	27.0 (7.8)	100.0	24.3 (8.7)	100.0	25.0 (5.9)	100.0	28.4 (5.4)	100.0	26.7 (5.6)	100.0

Note: Figures in brackets indicate sourcewise percentage compared to India.
*Indicates Anicutes, Bhandaras, Springs, Kuttai, Thangal, Small diversion networks and so on.

Source: Indian Agricultural Statistics, 1985-86—1989-90, Vol.I, Ministry of Agriculture, GOI. New Delhi.
CMIE, Agriculture, February 2004. Indian Agriculture in Brief 27th Edn. January 2000.
GOTN, Tamil Nadu-An Economic Appraisal, 1992-93, Evaluation and Applied Research Department, Chennai.
GOTN, Season and Crop Reports, Various Issues. Chennai.

On the whole, in TN, the contribution of different sources to NIA over a fifty-year period is not encouraging compared to the all India level. The NIA has increased only by 26 per cent over this period. It is merely a half a per cent increase per annum whereas at all India level the increase is more than two fold compared over the 1950s. In TN, if we take the decade 1970s as a mid point, then one can observe that from the 1950s to the 1970s irrigation from all sources has marginally increased. After that except wells and tube wells category, a steady decline in all sources is noticed including the total NIA of the state. This decline is more intense in the case of minor irrigation (MI) i.e. tanks and other sources than the canals. On the whole Table 3 indicates that the contribution of MI sources in TN is steadily declining but at all India level this is true only for tank irrigation but the contribution from 'other sources' is marginally increasing overtime.

Apart from the comparison between India and TN one can also view the extent of decline of tank irrigation in TN in relation to two other south Indian states such as Andhra Pradesh and Karnataka where tank irrigation is predominant even now. Here also we can compare the sources of irrigation between the states and India. If we analyse Table 4 the following points emerge.

At all India level the NSA has increased from 136 million hectares (M.ha.) to 142 M.ha. from 1960 1 to 1999 2000. During the same period the NIA has more than doubled from 25.5 M.ha. to 56.4 M.ha. This implies area, which is brought under cultivation moved up with a very slow pace compared to the area under irrigation. The percentage of NIA to NSA has doubled at all India level but across the southern states much variation is found. In absolute terms Karnataka shows a steady growth followed by Andhra Pradesh (AP) in terms of area under irrigation, whereas in TN the increase is verv slow. Across different sources of irrigation except wells the remaining sources did not show much progress at all India level whereas, across states Karnataka achieved impressive growth in all forms of irrigation except under tanks. In TN, only well irrigation showed some progress and in AP both canals and wells have achieved some progress.

Table 4 clearly brings out one point that both at all India level as well as across the three southern states the percentage of area under tanks has gradually declined. The decline is rapid in all the three states. More importantly there has been a marginal increase in the absolute area under 'other' category of MI sources both at all India level and across states except in TN. This implies that this source also needs to be developed extensively. It is also evident that TN is in a vulnerable position in terms of area irrigated by canals, tanks and other sources. Actually if we compare the progress of canal irrigation between TN and Karnataka from the early 1960s to the late 1990s the visible picture that one gets is that Karnataka's canal irrigation has steadily progressed from 2.8 lakh ha. to 9.5 lakh ha. (i.e., more than three fold increase) whereas in TN it has actually declined from 8.9 lakh ha. to 8.4 lakh ha. So, tables 3 and 4 show that TN's position

Table 4
Net Area Irrigated by Soures in India and three Southern States, 1960-1 to 2000

(Area in '000 hecs)

Source	*India*	*% of NIA*	*AP*	*% of NIA*	*TN*	*% of NIA*	*Karna-taka*	*% of NIA*	*4+6+8*	*Col. 4*	*Col. 6*	*Col. 8*	*Col. 10*
			1997-98 to 1999-2000								*as % of ol. 2*		
1	2	3	4	5	6	7	8	9	10	11	12	13	14
Canals	17598	31.2	1602	37.3	847	28.4	950	38.5	3398	9.1	4.8	5.4	19.3
Tanks	3114	5.5	675	15.7	666	22.4	247	10.0	1587	21.7	21.4	7.9	51.0
Wells	32557	57.6	1824	42.5	1448	48.6	916	37.1	4188	5.6	4.4	2.8	12.9
Others	3223	5.7	188	4.4	18	0.6	355	14.4	562	5.8	0.6	11.0	17.4
NIA	56491	100.0	4289	100.0	2979	100.0	2468	100.0	9736	7.6	5.3	4.4	17.2
NSA *	141971	39.8	10478	40.9	5560	53.6	10274	24.0	26312	7.4	3.9	7.2	18.5
Source						**1981-82 to 1985-86**							
Canals	16278	39.2	1783	49.1	840	33.0	657	41.5	3280	11.0	5.2	4.0	20.1
Tanks	3392	8.2	909	25.0	690	27.1	300	19.0	1899	26.8	20.3	8.8	56.0
Wells	19511	46.9	832	22.9	993	39.0	440	27.8	2265	4.3	5.1	2.3	11.6
Others	2376	5.7	108	3.0	21	0.8	186	11.7	315	4.5	0.9	7.8	13.3
NIA	41557	100.0	3632	100.0	2544	100.0	1583	100.0	7759	8.7	6.1	3.8	18.7
NSA *	141591	29.4	10941	33.2	5666	44.9	10415	15.2	27022	7.7	4.0	7.4	19.1
Source						**1960-61 to 1964-65**							
Canals	10791	42.3	1293	42.0	895	36.3	280	29.7	2468	12.0	8.3	2.6	22.9
Tanks	4668	18.3	1261	41.0	926	37.6	365	38.7	2552	27.0	19.8	7.8	54.7
Wells	7627	29.9	397	12.9	602	24.4	157	16.7	1156	5.2	7.9	2.1	15.2
Others	2454	9.6	127	4.1	43	1.7	140	14.9	310	5.2	1.8	5.7	12.6
NIA	25540	100.0	3075	100.0	2466	100.0	942	100.0	6483	12.0	9.7	3.7	25.4
NSA *	135908	18.8	11322	27.2	6039	40.8	10354	9.1	27715	8.3	4.4	7.6	20.4

Source: GOI, Indian Agricultural Statistics, Vol.I & II, 1992-93, New Delhi. CMIE, Agriculture, February 2004.
GOI. Indian Agriculture in Brief, (27th Edition), Directorate of Economics & Statistics, New Delhi. 2000.

Note: AP = Andhra Pradesh. TN = Tamil Nadu. NSA * denotes NIA as % of NSA.

in relation to surface sources of irrigation especially under tanks is not encouraging.

STATUS OF SMALL WATER BODIES

Before going to discuss the specific details of tanks in TN we have to obtain a clear picture of the contribution of tanks across states in India. Table 5 is constructed to serve this purpose. More than 99 per cent of NIA by tanks was concentrated only in 12 states of India and more than half is served only by the 4 southern states. This shows the importance of tank irrigation in the southern states. Further, the percentage of tank irrigated area to total NIA is between 10 and 22 per cent among the seven coastal states of India. This higher percentage share is mainly due to the locational advantage of the states, which facilitated to accommodate more number of smaller water bodies to hold the catchment run-off from both local as well as rivers.

Apart from tanks 'other sources' of irrigation at all India level also assumes importance. Because this category serves more or less equal extent of area as that of tanks. But its contribution to NIA varies across states. In the northern states such as Bihar, Madhya Pradesh and Uttar Pradesh area under 'other sources' outnumbered the tanks. Table 5 indicates that both the 'tanks' and 'other sources' of irrigation should be given equal importance for better development of MI sources across states in the country.

Now let us specifically discuss the contribution of tanks in TN. A glance at table 6 gives us the following points: In TN, 93% of NIA by tanks and 84% of total tanks are concentrated in 15 districts that are termed as *Intensive Tank Irrigation (ITI)* districts. Among these, NIA by tanks is more than 50% in 5 districts. In the remaining 13 less tank intensive districts of the state, the NIA by tanks in each is less than 7000 ha. and in all these districts the NIA by tanks to NSA is negligible between 0 and 5 %. Whereas in the ITI districts the same being relatively high and in two districts viz., Kancheepuram and Sivaganga, the percentage is more than 50.

In a few districts like Tiruchirapalli and Salem the mere existence of number of large tanks did not induce increase in the area under tank irrigation. In these two districts the average area irrigated per tank works out to 3 ha. and 5 ha. respectively and the state average being 13 ha. In contrast, the maximum NIA per tank is 38 ha. in Kancheepuram followed by 32 ha. in Ramanathapuram. It is evident that the Palar basin is the most tank intensive basin in TN followed by the Periyar Vaigai basin. Most importantly, all the ITI districts do have some form of river flows but not any perennial river. Hence the contribution of tanks is moderate in those districts. Since the Cauvery canals serve better in a few districts where the contribution of tanks is negligible. As indicated in the case of coastal states (where tanks are more intense than the inland states) the districts that are located in the coastal region of TN also come in the same category.

Table 5
Net Area Irrigated by Sources by Major States in India, 1997-8 to 1999-2000 (Area '000 ha)

Sl. No.	Name of State	NSA	Net area irrigated by								% of Col. 5 to Col.9
			Canals	Tanks	Wells	Other sources	Total 5+7	Total NIA	% of col.8 to col.3	% of col.8 to col.9	
1	2	3	4	5	6	7	8	9	10	11	12
1.	Andhra Pradesh	10478	1602	675	1824	188	863	4289	8.2	20.1	15.7
2.	Tamil Nadu	5560	847	666	1448	18	684	2979	12.3	23.0	22.4
3.	Maharashtra	17728	877	369	1799	89	458	3134	2.6	14.6	11.8
4.	Orissa	6082	949	305	836	0	305	2090	5.0	14.6	14.6
5.	West Bengal	5459	717	263	712	219	482	1911	8.8	25.2	13.8
6.	Karnataka	10274	950	247	916	355	602	2468	5.9	24.4	10.0
7.	Madhya Pradesh	15101	1513	184	3638	840	1025	6175	6.8	16.6	3.0
8.	Rajasthan	16219	1567	118	3762	51	168	5497	1.0	3.1	2.1
9.	Bihar	7426	1072	117	1794	641	758	3624	10.2	20.9	3.2
10.	Uttar Pradesh	16778	3084	91	8919	259	350	12353	2.1	2.8	0.7
11.	Kerala	2256	96	49	104	120	169	368	7.5	45.8	13.2
12.	Gujarat	9672	617	28	2395	21	49	3061	0.5	1.6	0.9
13.	Total of 12 states	123033	13889	3111	28147	2801	5913	47949	4.8	12.3	6.5
	% to Total NIA		29.0		58.7	5.8					
14.	India	141971	17598	3114	32557	3223	6336	56491	4.5	11.2	5.5
	13/14 (%)	86.7	78.9	99.9	86.5	86.9	93.3	352			

Source: CMIE, Agriculture, 2001, 2002 and 2004.
GOI. Indian Agriculture in Brief, (27th Edition), Directorate of Eco. & Statistics, New Delhi. 2000.

Table 6

Districtwise net sown, net irrigated and tank irrigated area in Tamil Nadu 2000-01 to 2002-03

(area in hectares)

Sl. No.	*Districts*	*2000-01 to 2002-03* *NSA*	*NIA by All Sources*	*NIA by Tanks*	*Col.5/ Col.4* 100*	*Col.5/ Col.3* 100*	*No.of Tanks during 2000-03 > 40 ha*	*No.of Tanks during 2000-03 < 40 ha*	*Total*	*Avg. NIA/ tank*	*Name of river basin*	
1	2	3	4	5	6	7	8	9	10	11	12	
1.	Ramanatha-puram	181864	67077	54570	81.4	30.0	477	1217	1694	32.2	PV + Gundar	
2.	Sivagangai	111817	79544	64481	81.1	57.7	651	4260	4911	13.1	Manimuthar + PV	
3.	Pudukkottai	144409	91652	63236	69.0	43.8	660	4791	5451	11.6	Vellar + Cauvery	4
4.	Kancheepuram	139219	126138	73137	58.0	52.5	709	1233	1942	37.7	Palar	2
5.	Kanniyakumari	80492	28559	15492	54.2	19.2	41	2582	2623	5.9	Kodayar	
6.	Virudhunagar	139483	57826	25419	44.0	18.2	290	707	997	25.5	Vaippar + Gundar	
7.	Thirunelveli	139015	95900	39801	41.5	28.6	373	1797	2170	18.3	Tambraparani + Chittar	7
8.	Thoothukudi	156338	35939	13449	37.4	8.6	107	527	634	21.2	Vaippar + Chittar	
9.	Thiruvallur	117796	91855	24903	27.1	21.1	780	906	1686	14.8	Ponnai + Cooum	1
10.	Villupuram	310013	179631	36922	20.6	11.9	988	1097	2085	17.7	Ponnaiyar + Gadilam	3
11.	Thiru-vannamalai	201885	124811	24114	19.3	11.9	604	1361	1965	12.3	Cheyyar	
12.	Madurai	135648	80224	15078	18.8	11.1	292	1995	2287	6.6	Periyar-Vaigai (PV)	5
13.	Dharmapuri	383679	117991	14358	12.2	3.7	216	2131	2347	6.1	Ponnaiyar + Chinnar	
14.	Vellore	195797	92866	8802	9.5	4.5	420	935	1355	6.5	Palar	
15.	Perambalur	200477	63045	5932	9.4	3.0	84	712	796	7.5	Cauvery	
16.	Theni	109460	56782	5289	9.3	4.8	55	151	206	25.7	Suruli + PV	
17.	Dindigul	227650	85413	7006	8.2	3.1	110	2120	2230	3.1	Kodavanar+Shanmukha	
18.	Tiruchirapalli	163308	94408	5042	5.3	3.1	115	1652	1767	2.9	Cauvery	
19.	Thanjavur	191737	162172	6693	4.1	3.5	130	298	428	15.6	Cauvery	

Contd...

Sl. No.	Districts	2000-01 to 2002-03 NSA	NIA by All Sources	NIA by Tanks	Col.5/ Col.4* 100	Col.5/ Col.3* 100	No.of Tanks during 2000-03 > 40 ha	No.of Tanks during 2000-03 < 40 ha	Total	Avg. NIA/ tank	Name of river basin
1	2	3	4	5	6	7	8	9	10	11	12
20.	Caddallore	222720	150272	5033	3.3	2.3	188	404	592	8.5	Cauvery + Vellar
21.	Namakkal	176790	59405	1723	2.9	1.0	67	192	259	6.7	Vellar
22.	Salem	243336	97646	2500	2.6	1.0	89	457	546	4.6	Vellar
23.	Karur	91983	46273	1021	2.2	1.1	18	248	266	3.8	Kodavanar+Shanmukha 8
24.	Coimbatore	310202	172733	1720	1.0	0.6	44	33	77	22.3	Aliyar+Bhavani+Noyyal
25.	Erode	281965	148253	163	0.1	0.1	21	31	52	3.1	Bhavani+Noyyal
26.	Thiruvarur	142903	137593	0	0.0	0.0	0	0	0	0.0	
27.	Nagapatinum	143824	121259	0	0.0	0.0	0	0	0	0.0	
28.	The Nilgiris	78227	877	0	0.0	0.0	0	0	0	0.0	
	Total	5022039	2666144	515884	19.3	10.3	7529	31837	39366	13:1	

Source: GOTN. Season and Crop Reports of Tamil Nadu, Various Issues. Dept. of Economics & Statistics. Chennai.

Where one can observe that all the ITI districts are located within the ambit of the coastal region.

On the whole the data on number of tanks in two size groups indicate that there are 39,366 tanks in TN. As indicated earlier, there is a serious doubt with respect to the total number of available tanks in TN. Because over a period of two decades more or less the same number of tanks are reported even though there has been wide fluctuations between the two categories of tanks. Hence, this should be verified and the actual hydraulic and other technical details need to be explored through census method. This can be initiated at least in a couple of tank intensive districts of the state.

It is well known that rainfall is the primary source for filling up of tanks, whether it is rainfed or river fed. Hence it is pertinent to look into this aspect of the relationship between rainfall and area under tank irrigation in TN over a period of 50 years. The rainfall data presented for 50-year period in Table 7 provides the following information. Over the five decades after independence, the decade 1960s is considered as the "golden period" of tank irrigation in terms of area irrigated by tanks. Only in this decade for seven years there was an increase in area under tank irrigation, which is more than 9 lakh hectares. During the decades of 1960s and 1990s, the decadal annual average rainfall is a maximum of 940 mm, which is slightly more than the decades of 1950s and 1970s. However, the decade 1980s was recorded a lowest rainfall of 856 mm which is 86 mm less than the decade of the 1990s. Even though average rainfall during these two decades are highly varying the NIA by tanks between these two decades are more or less the same (which is 6.2 lakh ha.). This indicates that over the period of 10 years available rainfall is not an important constraint for area under irrigation especially under tanks. But the main point that should be probed in detail is to what extent and how efficiently the farmers make use of the available rainwater from the catchment, in both categories of system and non-system tanks, for crop production by way of effective maintenance of irrigation structures and management of available water from the tanks.

Comparing the decades 1980s and 1990s, a higher average rainfall was recorded in the latter decade, which happened to be the highest among all decades since the 1950s. In the same manner it is observed that the decade of the 1980s had the lowest average rainfall compared to any other decade since the 1950s. Compared to the decades 1960s and 1970s the NIA served by tanks during the 1980s and 1990s declined considerably. One can ascertain that the decline of extent in the 1980s was mainly due to poor rainfall over many years but it is not certain the decade of 1990s where the average rainfall is the maximum compared to any other decade. Over the past two decades not even a single year has shown tank irrigated area of over 9 lakh ha. since 1977. This implies that other than rainfall there are other factors, which hamper the growth and sustainability of tank irrigation in TN. The exact reasons for these could be traceable only through intensive field observations.

Table 7
Size Class & NIA by Tanks and Rainfall in Tamil Nadu:1950-1 to 2000-1

Sl. No.	*Year*	*Number of Tanks*			*NIA*	*Rainfall in MM*				*Average*	
		>40ha	*<40ha*	*Total*	*by Tanks*	*June to Sep.*	*Oct. to Dec.*	*Jan. to May*	*June to May*	*10*	*5 years*
1	2	3	4	5	6	7	8	9	10	11	12
Normal Rainfall ->	As per 2002-03 SCR					329.5	468.1	166.5	964.1		
Normal Rainfall ->	51 Years following data					321.0	446.2	150.5	917.8		
1.	1950-51	7525	15459	22984	545.4	283	350	215	847.5		
2.	1951-52	6435	17380	23815	635.7	335	325	158	817.5		
3.	1952-53	6793	18378	25171	617.8	198	368	183	747.5		
4.	1953-54	6652	18512	25164	795.5	390	405	263	1058.0		
5.	1954-55	6628	18613	25241	861.9	335	415	300	1049.8		904
6.	1955-56	7276	17798	25074	794.4	340	420	96	855.8		
7.	1956-57	7327	20117	27444	884.3	395	450	153	997.8		
8.	1957-58	6994	21366	28360	847.4	330	500	230	1060.0		
9.	1958-59	7459	21013	28472	840.1	343	401	172	916.0		
10.	1959-60	6986	22612	29598	832.8	342	459	168	969.0	932	960
11.	1960-61	7150	22753	29903	936.4	349	580	195	1124.0		
12.	1961-62	7030	27437	34467	939.5	371	309	187	867.0		
13.	1962-63	7220	27703	34923	946.2	325	395	212	932.0		
14.	1963-64	7562	29246	36808	914.9	315	506	87	908.0		
15.	1964-65	7842	28323	36165	892.2	347	406	106	859.0		938
16.	1965-66	8611	26730	35341	902.5	315	441	115	871.0		
17.	1966-67	8183	28187	36370	965.8	416	606	139	1161.0		
18.	1967-68	8452	28180	36632	989.9	285	496	179	960.0		
19.	1968-69	8442	28193	36635	750.5	271	312	100	683.0		
20.	1969-70	8726	27019	35745	879.2	238	613	186	1037.0	940	942
21.	1970-71	8544	27278	35822	897.9	318	420	180	918.0		
22.	1971-72	9358	28056	37414	923.5	323	489	147	959.0		
23.	1972-73	9286	27917	37203	949.5	304	608	79	991.0		
24.	1973-74	8093	29776	37869	928.5	333	407	102	842.0		
25.	1974-75	8427	29547	37974	593.9	326	178	143	647.0		871
26.	1975-76	7386	30593	37979	749.8	420	340	97	857.0		
27.	1976-77	7412	30923	38335	800.0	314	440	187	941.0		
28.	1977-78	7480	31461	38941	909.9	333	682	109	1124.0		
29.	1978-79	7402	30795	38197	841.9	261	582	107	950.0		
30.	1979-80	7398	30896	38294	896.2	361	605	125	1091.0	932	993
31.	1980-81	7399	30895	38294	590.2	196	337	136	669.0		
32.	1981-82	7408	30896	38304	738.4	406	453	97	956.0		
33.	1982-83	7116	30828	37944	516.6	217	352	94	663.0		
34.	1983-84	7125	30591	37716	807.1	399	484	340	1223.0		
35.	1984-85	7191	30438	37629	715.0	331	300	160	791.0		860
36.	1985-86	7224	30421	37645	671.7	382	377	192	951.0		
37.	1986-87	7200	30749	37949	509.6	272	331	98	701.0		
38.	1987-88	7280	30949	38229	609.9	271	525	187	983.0		
39.	1988-89	7282	31067	38349	479.0	376	218	115	709.0		
40.	1989-90	7281	31202	38483	522.7	349	341	227	917.0	856	852

Contd...

Sl. No.	*Year*	*Number of Tanks*			*NIA by Tanks*	*Rainfall in MM*				*Average*	
		>40ha	*<40ha*	*Total*		*June to Sep.*	*Oct. to Dec.*	*Jan. to May*	*June to May*	*10*	*5 years*
1	2	3	4	5	6	7	8	9	10	11	12
41.	1990-91	7281	31202	38483	530.9	239	373	103	715.0		
42.	1991-92	7299	31453	38752	577.0	332	485	82	899.0		
43.	1992-93	7176	31687	38863	628.8	316	477	69	862.0		
44.	1993-94	7164	31678	38842	668.1	305	710	78	1093.0		
45.	1994-95	7169	31694	38863	674.3	224	479	231	933.8		901
46.	1995-96	7423	31283	38706	512.3	348	248	68	663.7		
47.	1996-97	7476	31527	39003	623.5	455	541	125	1121.2		
48.	1997-98	7483	31520	39003	675.5	286	782	84	1152.2		
49.	1998-99	7522	31776	39298	689.7	340	602	138	1080.0		
50.	1999-00	10070	31878	41948	633.1	200	500	197	896.8	942	983
51.	2000-01	7529	31837	39366	588.6	315	336	135	785.3		

Source: GOTN, Season and Crop Reports, Various Issues, Dept. of Economics and Statistics.

Whenever October December rainfall is more than 600 mm, the area served by tanks in that particular year was more than 9 lakh ha. This point holds good for five years during the decades 1960s and 1970s. But, the rainfall in the two other decades of 1950s and 1980s did not reach this figure. However, in the 1990s this point seems misleading, where in three years 1993 4, 1997 8 and 1998-99 the October December rainfall was 710 mm, 782 mm and 602 mm respectively. But, the extent of area served by tanks was only less than 7 lakh hectares, which is more or less same to that of the other normal rainfall years in that decade. This point may indicate that the area served by tanks had been encroached considerably or farmers' interest for cultivation under tanks may be declined over the past two decades. This needs to be studied in depth through micro level surveys.

October December rainfall is a decisive factor for agricultural prosperity in a year in TN. If Oct. Dec. rainfall is >400 mm then the year will be considered as normal/above normal year. If June September rainfall is >300 mm coupled with the Oct Dec. rainfall is also >400 mm then the year will be considered definitely as a normal/above normal year. Even though the rainfall norm set for June September is >300 mm and for Oct Dec. is >400 mm it is not invariant within this two seasonal periods. That is in one period if the rainfall is too high and the other monsoon period is slightly lower than the norm prescribed above may also fit well in our prediction in the normal/above normal rainfall pattern in TN. In other words, we can take the average of 700 mm rainfall as the basic criteria for both the seasons to arrive at the prediction of normal and above normal year.

Compared to 1967-68 where the area irrigated by tanks was 9.9 lakh ha. over a period of 33 years up to 2000-01 some 920 tanks under the >40 ha. category had disappeared. However, the smaller size tanks had increased to about 3660 during the same period. But the area irrigated by them decreased considerably. This raises a question whether the larger tanks are converted into smaller tanks due to encroachments. Between the 1960s and 1990s, the area under tanks decreased by 3 lakh hectares. This has happened only from 1984 5. Since then the maximum area served by the tanks was only 6.9 lakh ha. which happened to be in 1998 9, a good rainfall year. In both the 1960s and 1990s the average annual rainfall was more than 940 mm. But the average area irrigated by tanks in the latter decade is much less, which is only 6.3 lakh ha. Whereas the same being 9.1 lakh ha. during 1960s. The difference is nearly 3 lakh ha.

Table 7 amply indicates that although rainfall is an important factor, which decides the extent of area brought under tank irrigation, there are several other factors (which may be termed as physical, technical and institutional), which stand in the way of progress of tank irrigation in the state. Hence, this is an important issue, which needs to be probed in detail through field surveys to find out exact reasons for the decline of area under tanks in TN. Some other details could also be drawn while analysing table 7, and this may be represented through a graph, which shows the following:

1. There has not been any persistent decrease in rainfall rather it merely fluctuates over the years.
2. Above normal rainfall (>950mm) has occurred in many years. This is 22 out of 51 years between 1950 1 to 2000 1.
3. Average/marginal deficit rainfall (950 850mm) years are 15 out of 51 years.
4. Deficit rainfall (850 650) years are 14 out of 51 years.
5. Minimum rainfall is 647mm (1974 5) and the maximum is 1223mm (1983 4) over the 51 year period.
6. The relationship between rainfall and NIA by tanks is clear. That is, whenever the rainfall increases the NIA by tanks also increases and vice versa. But, only in one period (1960s) even though the rainfall was relatively low the NIA by tanks was relatively more consecutively for 5 years. But after that nowhere one can see such a pattern.

Apart from tanks, in most of the villages in India there are other smaller types of water bodies available. They are generally referred to as Kulam, Kuttai, Kasam and Tangal. (other common types are known as Anicuts, Pickups, Bhandaras, Springs, small diversion networks and so on). If all these sources are filled up during the monsoon period it would help to recharge the groundwater to a considerable extent. Apart from that these sources will also serve several other local purposes. To provide an example of this one can refer to Table 8. In one of the villages served by Kaveripakkam tank, there were 21 small water bodies found in the year 1913. But the existence of these sources needs to be verified now.

Table 8
Names of Small Water Bodies in the Villages Served by Kaveripakkam Tank, 1913.

Sl. No.	*Name of Village*	*Name of Water Bodies*	*No. of SWBs*
1.	Kaveripakkam	1. Tamara Kuttai.	1
2.	Kondapuram	1. Kandappar Kulam. 2. Nallathanni Kulam.	2
3.	Sirukarumbur	1. Krisnan Kasam. 2. Amarapuram Kasam. 3. Ayapadi Kasam. 4. Ocheri Kasam.	4
4.	Athipattu	1. Sirukarumbur Kasa Kalvay. 2. Meclean Kalvay. 3. Mamandur Kasam. 4. Vilagam Kasa Kalvay.	4
5.	Iralacheri	1. Siteri Tangal. 2. Isvaran Kulam. 3. Setti Kulam. 4. Nallathanni Kulam. 5. Pillaiyar Kuttai. 6. Nanthi-Setti Kuttai. 7. Paracheri Kuttai. 8. Thandu Kuttai. 9. Periadevi Kuttai. 10. Karai Kuttai. 11. Sundu Kuttai.	11
6.	Cheri	1. Nallathanni Kulam. 2. Alli Kulam. 3. Isvarankoil Kulam. 4. Pillaipettai Kuttai. 5. Anniyamman Kuttai.	5
7.	Kattalai	Wholly a wetland village located adjacent to the foot of KPT.	0
8.	Thuraiperum. pakkam	1. Pakkiri Kuttai. 2. Vannan Kuttai. 3. Para Vannan Kuttai. 4. Amman Kuttai. 5. Talattiyan Kuttai. 6. Pattan Kuttai. 7. Ayanan Kuttai.	7
9.	Maganipattu	1. Divittiyan kulam. 2. Toppu Kuttai. 3. Mangkuttai. 4. Thenginaru Kuttai. 5. Jambu Kuttai. 6. Sangara Kuttai. 7. Elaval Kuttai.	7
10.	Alapakkam	1. Pudu Tangal. 2. Isvaran Kulam. 3. Govindan Kulam. 4. Nallathanni Kulam. 5. Tamara Kulam. 6. Paracheri Kuttai. 7. Nirali Kuttai. 8. Vella Kuttai. 9.Pidari Kuttai. 10.Thulukkan Kuttai. 11. Kosavan Kuttai. 12. Mel Kuttai. 13.Munuperumal Kuttai. 14. Andar Kuttai. 15. Ellaitarisu Kuttai. 16. Parayan Kuttai. 17. Ramadas Kuttai. 18. Sudukattu Kuttai. 19. Petta Kuttai. 20. Lingam Kuttai. 21. Muttivettan Kuttai.	21
11.	Puduppattu	1. Sitteri Tangal. 2. Kurutu Anichan Tangal. 3. Pudupattu Tangal. 4. Vellalan Kulam. 5. Pillaiyar Kulam. 6. Govindan Kulam. 7. Kosavan Kuttai. 8. Enavayan Kuttai. 9. Mungil Kuttai. 10. Pandaram Kuttai.	10
12.	Panniyur	1. Nallathanni Kulam. 2 Palani Mudali Kuttai. 3. Ayakaran Kuttai.	3
	Total		75

Source: Descriptive Memoirs of Villages Served by Kaveripakkam Tank, 1913.

Annual Rainfall and NIA by Tanks in TN, 1950-51 to 2000-01

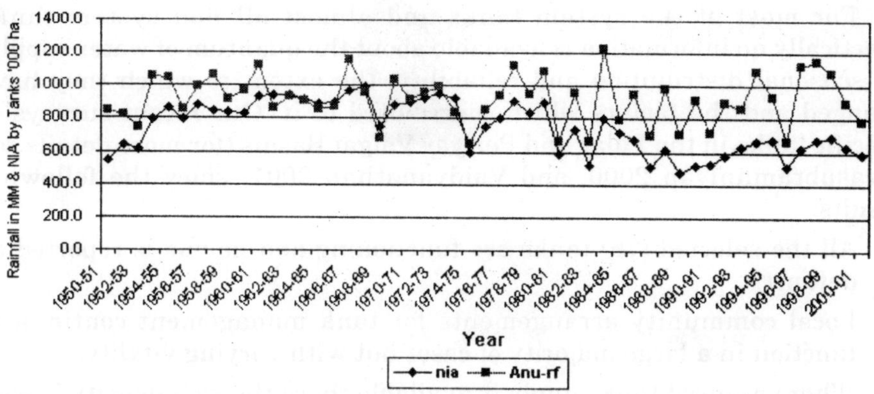

Seasonal Rainfall and NIA by Tanks, 1950-51 to 2000-01

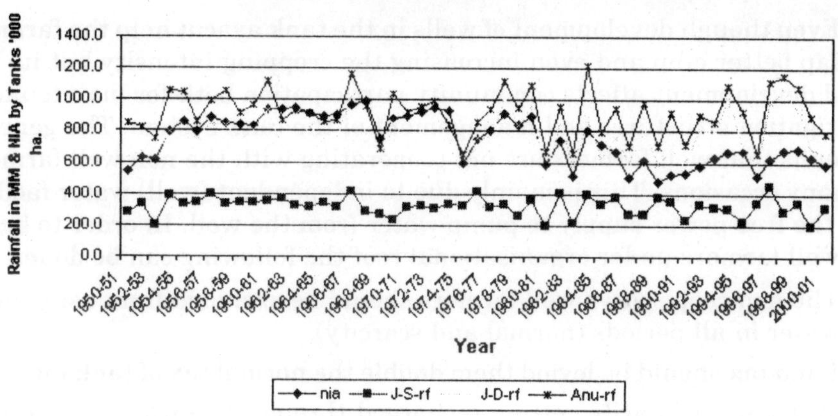

V

MAJOR PROBLEMS REPORTED IN TANK IRRIGATION STUDIES

Most of the available official statistics of tanks are inadequate, inconsistent and confusing. There is not even an authenticated count of the number of tanks and the actual extent of area irrigated by them (Vaidyanathan 2001). For instance, most of the available regional field studies on tanks indicate that not much reduction was taken place in effective ayacut over time; some in fact reported an increase due to the conversion of adjacent rainfed lands into the ayacut. But the point is at the macro level, both all India and TN, we cannot see the full extent of irrigated ayacut under tanks even when there is good rainfall. This is more so for the past two decades. This raises doubts about the way in which the irrigated area by tanks or other sources are recorded by the Village Administrative Officials. If the recorded/ reported data are correct then one can definitely say that siltation and reduced inflow into tanks significantly reduced the area actually irrigated by them (see also Sakthivadivel et.al 2004).

For most of the system tanks and almost all non system tanks practically no information is available about the quantum of water supply, its seasonal distribution and reliability, the extent to which they have changed and the factors which contributed to it. Our recent surveys of specific tanks in the Palar and Periyar Vaigai Basins (for more details see Sivasubramaniyan 2000; and Vaidyanathan 2001) show the following results:

1. All the selected (76) tanks are functioning and no one is reported as defunct.
2. Local community arrangements for tank management continue to function in a large majority of cases but with varying vitality.
3. Where assured tank supply is available there the well density is less, which leads to active community participation for tank management in a majority of cases and vice versa.

Even though development of wells in the tank ayacut help the farmers to reap better crop and even increasing the cropping intensity but its lop sided development affects community participation both for maintenance of irrigation structures and management of the tank system. The general notion is that well farmers are not cooperating with the non well farmers in many occasions. This is mainly due to independent (well) water facility and the free power supply to pump water from the well. In order to bring the well farmers under organisational roof the following can be done:

1. The non cooperative well farmers should be excluded from using tank water in all periods (normal and scarcity).
2. Land tax should be levied them double the normal tax of tank farmers.

Wells are primarily getting recharged through two ways: (i) through rainfall and (ii) through surface sources of irrigation such as canals and tanks. This makes the point that development of well irrigation implies development of surface sources of irrigation. In other words, if canals and tanks are not effectively developed the well water supply will not be prosperous. This means that if huge investments on major and medium irrigation projects are made the indirect benefits are immediately realised by well farmers who are situated in the vicinity of these projects.

CRITICAL FACTORS AFFECTING TANK MANAGEMENT

Apart from the general notion, the three important sets of factors that affect the functioning of tank irrigation management are: Physicai, Institutional and Technical. The Physical factors are: location and sources of supply; siltation and growth of weeds in the tank and in supply channels; encroachments in supply channels, foreshore area, tank beds and in the catchment area. If all these physical factors are in good condition then every tank can hold water to its full capacity.

The institutional problems are more severe in recent decades, which cannot be overcome unless a strong organisational set up in each tank is made mandatory. The most important problematic factors of this category are: social (caste and class structure); economic (land holding pattern) and demographic (population pressure on land). Apart from this, the technical factors such as conditions of water supply; drainage conditions, soil quality, the spread of well irrigation in the ayacut also lead to problems in the effective functioning of institutions. Most importantly, all these factors do not act in isolation and there is strong inter connection among them, which complicates the smooth functioning of institutions under tank irrigation all over the country (for a more detailed discussion on these aspects see Janakarajan 1993, and Sivasubramaniyan 1995).

VI

POLICY RECOMMENDATIONS FOR IMPROVEMENT AND RESTORATION OF TANKS

Although in recent decades–especially after the 1960s- our country has not experienced any famine (severe shortage of food) but the drought (temporary or a prolonged shortfall of rains) situations are prevailing in many parts of India. The latter phenomena happen mainly because of 'climatic factors' that are beyond human control. Even then, in some years, the country is receiving abundance of rainfall within a short span of time during the monsoon period and it is not properly utilised currently for beneficial purposes. This is one of the basic reasons for the declining trend in the area served by tanks. To make the tank irrigation system prosperous the following measures should be undertaken through the Central/State governments and the beneficiaries also to be involved in the process effectively.

First of all to find out the possibilities of converting local ponds/ kuttais into full fledged tanks and also to identify the possibilities of converting non system tanks into system tanks, which would enable to fully utilise the available surface water in a most beneficial manner. To survey the existence of all tanks and their present position in the state as well as in the country and a proper mapping of the tanks is a crucial need.

The possibility of studying the effectiveness of tanks in terms of ground water banking is important. That is, the quantum withdrawal and quantum recharged in a given year or for a continuous period, say 10 years. In TN at present there are 15 ITI districts identified. For each district at least 2 high powered bulldozers are to be given by the government and if they are directed and operated by the agricultural engineers, within 3 years we can restore the lost capacity or the original capacity of almost 90% of the existing tanks in TN. The amount spent for this operation can be recovered through increased agricultural productivity within a short span of time.

We can use the Geographic (Satellite) Imagery System (GIS) for this tank improvement programme effectively. Currently, the Indian Space

Research Organisation (ISRO) was the field leader in the GIS and India was one of the advanced countries, which had succeeded in using the system for developmental communication to reach rural poor. Further it was also possible to predict agricultural production by the remote sensing satellite. So, technology-wise our country is bright but the question is to what extent and how it is effectively utilised for the required purpose.

Our research findings would serve no purpose unless they are socially relevant. As M.S. Swaminathan has rightly pointed out, in many of his talks, that any crisis in agriculture would have an adverse impact on the country's economy. Any steps other than improving the purchasing power of rural people, who are mostly farmers, would only be a mere fire fighting exercise. This point is more relevant here because a majority of tank farmers are small and marginal farmers in the country.

The Government should immediately initiate a detailed investigation of basic information for all tanks in the state as well as the country as a whole. Further, a periodic desiltation and strengthening of tank infrastructure should be undertaken as routine process, which should not be withheld on any account at any time. Then only the restored capacity can be sustained. Also, legislations should be enacted (for eviction of encroachers and punishing illegal water users etc.), legal measures to be tightened and powers must be given to local Revenue and PWD authorities along with the local tank organisation to undertake maintenance and management of the tank system effectively. Care must be taken that these legal measures should not affect inter departmental functions and these laws should actually help the authorities concerned to improve tank efficiency.

In order to strengthen water use efficiency and improvements in tank performance the state agencies such as PWD and Revenue authorities as well as the NGOs should be involved to focus their attention more on educating farmers to understand the necessity of conserving/improving tank water supply and its use in an effective manner and avoiding wastage of water at each stage (i.e. at augmentation, storage and use).

The government's effort should be, wherever possible, to link the southern peninsular rivers first and then these should be interconnected to bigger tanks, which must be given priority in the planning process. The aim of interlinking of rivers (wherever possible) should be to stop flooding during the monsoon months and also to stop water reaching the Sea from all the rivers. What we need to do this is a conscious effort to explore and exploit this possibility. This naturally requires planners to view canals and tanks in a given region as complementary and not as separate entities. If it is done the following benefits can be availed:

1. Flooding can be reduced considerably.
2. Severe drought situation can be averted.
3. Groundwater recharge could be increased.
4. Additional area would be brought under plough.

5. Dry farming will be converted into wet farming that automatically enhances the productivity.
6. High fluctuations in foodgrains production could be averted.
7. This would increase the employment potential and also enhance the standard of living of poor farmers through increased income.

It is pertinent to note the observations made by Vaidyanathan as: "the programmes for tank irrigation in recent decades have been inadequate in scale, misconceived in design, poor in implementation and dubious in their impact. Irrigation development has been dominated by large surface works and groundwater exploitation. They have absorbed most of the investment in this sector. Allocations for small surface irrigation, of which tanks are the most important, have been quite meagre by comparison and also in relation to the needs of this segment." (Vaidyanathan 2001: 157). Hence, prime importance to be given in the Plan investments for the betterment of tank irrigation.

VII

CONCLUSIONS

In recent years most Indian states are increasingly facing water scarcity not only during drought years but also in normal rainfall years. Water disputes are frequently referred almost in all south Indian states. In Tamil Nadu all available surface (river) flows are mostly utilised and there is no scope for expansion of canal irrigation system. For the past five decades, the canal irrigated area in Tamil Nadu is stagnant which is around eight lakh hectares. Whereas in the neighboring state of Karnataka the same being gradually and steadily increasing over the same period; and there is also bright scope for further expansion of this source in that state. Under this circumstance in future, it is very difficult to share their river waters fairly with Tamil Nadu. The Kerala river supplies to Tamil Nadu are also gradually dwindling due to want of water for power generation in Kerala. Next, groundwater utilisation in Tamil Nadu has also reached its limit. Most of the districts aquifers are already marked in the dark region and there is no scope for further expansion of this source. So, in all counts expansion of irrigated agriculture in Tamil Nadu is in stark. The same may also be the case for other states in the near future. At this juncture, the only solution that is apparently available is immediate revival of Small Water Bodies in all states.

The necessity for expansion and revival of tank irrigation arises due to the following counts. After green revolution irrigated agriculture mostly depends its source of water from well irrigation. Water supply available from Independent wells are much limited. Only supplemental wells located in the canal and tank commands provide adequate and reliable supply. Whenever canals/tanks did not get adequate supply, the wells located in the vicinity of the surface sources also get poor recharge and the independent wells get almost negligible recharge through poor rains. This clearly point out the necessity for the revival of tanks. Wherever tank irrigation systems are available there the well water recharge is appreciably more than the independent wells' recharge through rainfall. Due to this fact, the right way to resolve our, ever expanding, water problem is to provide adequate attention on watershed development to conserve rainwater, revival of all types of smaller water bodies to recharge groundwater, optimal use of available supplies form the major and medium irrigation projects, involving the users at large through formation of water users' associations to economic use of water and avoidance of waste. If all these practices are undertaken step by step by the states and encouragement as well as adequate resources are earmarked through Central Plan investments our future water problems will be solved beyond doubt.

NOTES

1. For an overview of the factors responsible for variations in nature of irrigation works and irrigation organisations see Vaidyanathan 1985.
2. The available literature is reviewed in Vaidyanathan 1985.
3. The experiments, their achievements and the problems encountered in different parts of the world have been documented in papers presented at the "International Conference on Irrigation Management Transfer" (IIMI 1994); for a brief review of the conference proceedings see Vaidyanathan 1994b.
4. Geertz 1967; Fernea 1970; Gray 1963; Downing and Gibson 1974; Coward 1980; Vaidyanathan 1985; Kelly 1980; Leach 1961.
5. Chambers 1977, 1988; Narayana 1982; Vaidyanathan and Janakarajan 1989; Sengupta 1980,1982, 1984, 1991, 1993; Ramanathan and Ghose 1994; Rajagopal 1991.

REFERENCES

Chambers, Robert (1977), Men and Water: The Organisation and Operation of Irrigation. In B.H. Farmer. (ed.) *Green Revolution? Technology and Change in Rice-Growing Areas of Tamil Nadu and Sri Lanka*. pp. 340-61. London: The Macmillan Press Ltd.

——— (1988), *Managing Canal Irrigation: Practical Analysis From South India*. Cambridge: Cambridge University Press.

Coward, Jr. E.W. (1980), *Irrigation Development: Institutional and Organisational Issues* In E.W. Coward (ed.). Irrigation and Agricultural Development in Asia: Perspectives from the Social Sciences. pp. 15-27. New York: Cornell University Press.

Dikshit, G.S., G.R. Kuppuswamy and S.K. Mohan (1994), *Tank Irrigation in Karnataka: A Historical Survey*. Bangalore: Gandhi Sahitya Sangha.

Downing, Theodore E. and Gibson, McGuire (eds.) (1974), *Irrigation Impact on Society*. Tucson: University of Arisona Press.

Fernea, Robert (1970), *Shaykh and Effendi: Changing Patterns of Authority Among the El Shabana of South Iraq*. Cambridge: Harvard University Press.

Geertz, C. (1967), *Organisation of the Balinese Subak*. In E.W. Coward. (ed.). Irrigation and Agricultural Development in Asia: Perspectives from the Social Sciences. pp. 70-90. New York: Cornell University Press.

Gray, R.F. (1963), *The Sanjo of Tansanyika: An Anthropological Study of an Irrigation-based Society*. London: Oxford University Press.

Hunt, Eva and Hunt, Robert, C. (1976), Canal Irrigation and Local Social Organisation. *Current Anthropology*. 17(3): 125-34.

IIMI. (International Irrigation management Institute) (1994), International Conference on Irrigation Management Transfer. (Draft Conference Papers). Wuhan. China.

Janakarajan, S. (1993), In Search of Tanks: Some Hidden Facts. *Economic and Political Weekly*. 28 (26): A 53-60.

Kelly, W.W. (1980), *Water Control in an Agrarian State: Irrigation Organisation in a Japanese River Basin*. Ph.D. thesis. London: University Microfilms International.

Krishnaswami, S.Y. (1947), *Rural Problems in Madras*. Madras. Government Press.

Leach, E.R. (1961), *Pul Eliya: A Village in Ceylon*. Cambridge: Cambridge University Press.

Ludden, David (1978), *Agrarian Organisation in Tinnevelly Dis-trict, 800 to 1900 A.D.* Ph.D. Thesis. University of Pennsylvania. University Microfilms International.

Maass, A. and R. Anderson (1978). ... *and the Desert Shall Re-joice: Conflict, Growth, and Justice in Arid Environment*. Cam-bridge: MIT Press.

Millon, Rene (1962), Variations in Social Responses to the Prac-tice of Irrigated Agriculture. In Richard B. Woodbury. (ed.). *Civilisation in Desert Lands*. USA: Utah State University.

Narayana, D., V.C.V. Ratnam and K.Narayanan Nair (1982), An Approach to Study of Irrigation: Case of Kanyakumari District. Review of Agriculture. *Economic and Political Weekly*. 17 (39) A 85-102.

Ostrom, Elinor (1992), *Crafting Institutions for Self-Governing Irrigation Systems*. San Francisco: ICS Press.

Rajagopal, A. (1991), *Water Management in Agriculture With Special Reference to Irrigation Institutions: A Comparative Study of Canal and Tank Commands in Cauvery Basin, Tamilnadu*. Ph.D. the-sis. New Delhi: Jawaharlal Nehru University.

Raju, Sarada (1941), *Economic Conditions in the Madras Presidency. 1800-1850*. Madras: University of Madras.

Ramanathan, S. (n.d.). *Controversy Around Kudimaramat in Madras Presidency*, mimeo.

Ramanathan, S. and Ghose (1994), Irrigation Management Turnover: A User's Perspective: The Case of the Indira Gandhi Canal, Rajasthan. India. In *IIMI and Wuhan University of Hydraulic and Electrical Engineering*. Wuhan. China. Volume I : 173-84.

Reddy, D. Narasimha (1990), Decline of Traditional Water Harvesting Systems: Tanks in the Drought Prone Areas of Andhra Pradesh. Seminar on *Traditional Water Harvesting Systems in India*. October 9-11. New Delhi. Centre for Science and Environment.

Sakthivadivel, R., P. Gomathinayagam and Tushaar Shah (2004), Rejuvenating Irrigation Tanks through Local Institutions. Economic and Political Weekly, 39 (31) 3521-26.

Sengupta, Nirmal (1980), The Indigenous Irrigation Organisation in South Bihar. *The Indian Economic and Social History Review*. 17 (2): 157-89.

——(1982), *Tank Irrigation in Gangetic Bihar*. (mimeo). Patna: A.N. Sinha Institute of Social Studies.

——(1984), A Political Economic Study of an Indigenous Irrigation System. Ph.D. Thesis. Calcutta: Indian Statistical Institute.

——(1985), Irrigation: Traditional vs Modern. *Economic and Political Weekly*. 20 (45-47): 1919-38.

——(1991), *Managing Common Property: Irrigation in India and the Philippines*. New Delhi: Sage Publications.

——(1993)m *User-Friendly Irrigation Designs*. New Delhi: Sage Publications.

Sivasubramaniyan, K. (1985), *Economics of Water Management Under Canal Irrigation*. M. Phil. Dissertation in Economics. Tiruchira-palli: Bharathidasan University.

——(1995), Irrigation Institutions in Two Large Multi-Village Tanks of Tamil Nadu: Structure, Functioning and Impact. Ph.D. *thesis*. MIDS. University of Madras.

——(1997), Irrigation Management under Different Conditions: A Study of Large Tank Systems in Tamil Nadu in *Water Nepal*. Vol. 5. No. 2. July-December.

——(1998a). Maintenance of Tank Irrigation Systems: Study of Two System Tanks in Tamil Nadu in *Review of Development & Change*. Vol. III. No. 2. July-December.

——(1998b), Irrigation under the Palar Anicut System: A Case Study from Tamil Nadu in *Artha Vijnana*. Vol. XL. No. 4. December.

——(2000a), Tank Irrigation under the Periyar Vaigai Basin: A Historical Perspective in Institute for Social and Economic Change Journal. *Journal of Social and Economic Development*: Vol. III. No. 2. July-December.

——(2000b), Water Management Traditions under Major Tanks: with Special Reference to Mamulnamas. MIDS Journal. *Review of Development & Change*. Vol. V. No. 2. July-December.

Steward, Julian H. (1949), Cultural Causality and Law. Initiation of a Research Trend: Wittfogel's Irrigation Hypothesis. In *Socie-ty and History Essays in Honour of K.A. Wittfogel*. G.L. Ulman (ed.) New York: Morton.

Tang, Shui Yan (1992), *Institutions and Collective Action Self-Governance in India*. San Francisco: ICS Press.

Vaidyanathan, A. (1984), *Organisation and Management of Water Control in China*. V.R.F. Series No. 113. Tokyo: Institute of Developing Economies.

——(1985), Water Control Institutions and Agriculture: A Com-parative Perspective. *Indian Economic Review*. 20 (1): 25- 83.

——(1991), Critical Issues Facing Indian Irrigation. In Ruth Meinzendick and Mark Svendsen (eds.). *Future Directions for Indian Irrigation: Research and Policy Issues*. International Food Policy Research Institute. Washington. D.C.

—— (1992), *Strategy for Development of Tank Irrigation*. Madras: Madras Institute of Development Studies.

—— (1994a), Second India Studies Revisited: Food, Agriculture and Water. Madras: Madras Institute of Development Studies.

—— (1994b), Transferring Irrigation Management to Farmers. *Economic and Political Weekly*. 29 (47): 2965-67.

Vaidyanathan A. (ed.) (2001), *Tanks of South India*. Centre for Science and Environment. New Delhi.

Vaidyanathan, A and S. Janakarajan (1989), *Management of Irriga-tion and its Effect on Productivity: A Study of Two Surface Irrigation Systems in Tamilnadu*. mimeo. Madras: Madras Insti-tute of Development Studies.

Vaidyanathan, A and K. Sivasubramaniyan (2001), Tank Irrigation in South India: An Overview. In A. Vaidyanathan (ed.). *Tanks of South India*. Centre for Science and Environment. New Delhi.

Wade, Robert (1987), *Village Republics: Economic Conditions for Collective Action in South India*. Cambridge: Cambridge University Press.

Vander Velde, E.J. (1980), Local Consequences of a Large-Scale Irrigation System in India. In E.W. Coward. (ed.). *Irrigation and Agricultural Development in Asia*. pp. 291-328. New York: Cornell University Press.

Wittfogel, Karl. A. (1957), *Oriental Despotism: A Comparative Study of Total Power*. New Haven: Yale University Press.

Government of India (1989), *Central Board of Irrigation and Power*. Statistics of Minor Irrigation Development. New Delhi. Government Press.

——(1992), Planning Commission. *Report of the Committee on Pricing of Irrigation Water*. New Delhi: Government Press.

Government of Tamil Nadu (1979), *Report of Tank Irrigation Modernisation Project*. Mimeo. Madras. Government Press.

28

SUSTAINABILITY OF AGRICULTURAL PRACTICES IN WEST BENGAL —A CASE STUDY OF MALDA DISTRICT

K.K. Bagchi & Swajan Kumar Dey

INTRODUCTION

Indian agriculture has under gone several changes over the decades and centuries that have passed. Since the time of independence, a radical transformation has occurred in several facets of Indian agriculture due to planned investments that have resulted in noteworthy economic progress.

Indian agricultural development can be looked at from two different perspectives: institutional and technological. The institutional approach mainly typified by land reforms and changing agrarian relations, was the strategy-adopted way back in mid-fifties. The technological approach to the agricultural development came around the mid sixties and made decisive impact on agricultural production and productivity, most ostensibly in the regions that had the basic pre-requisites for switching over to new technology. The switch over from a food deficit to a food surplus economy was plainly the outcome of technology institutional human interface. It also deserves to be underlined that the green revolution technology led to sizeable increased in returns to land, and expanding needs for farm inputs, and milling and marketing which stimulated expansion of the rural non-farm economy. But, in recent years this interface seems to be falling apart. Technology that heralded the process of green revolution during late nineteen sixties started showing sings of wariness during the nineties. This happened for a number of reasons that are largely of our own making. Soil degradation, deforestation, water scarcity and contamination, bio-diversity loss and climate changes are increasingly at risk. The technology weariness reflects in itself, in stagnating and fillip yield rates for a number of crops in a number of regions. It may look unpleasant, yet an objective assessment of the history of the green revolution technology would point to its un-enlightened if not unwise use, guided by short-term gains of the rate of resource depletion, most markedly for land and water that has been occurring in different parts of Indian agriculture, in the name of improved agricultural technology. The unsustainable became the future growth of agriculture.

The increasing degree of technology sophistications, the strategic importance of market intelligence, and the critical importance of input-output process and price parities, are putting the decision making times. With every passing day, their difficulties are getting compounded. Today, a typical Indian farmers has not only to look back at the unwise pattern of resource use by his predecessor, and devise do's and don't for himself, but has also contended with numerous challenges that visit him on the technology and marketing fronts under the open economic regime.

For sustained agricultural development Indian agriculture must rely upon improved agricultural technology and this technology has been incorporated into the farming system of India in a significant way. To achieve all the planned targets an appropriate technology has to be evolved so that agricultural production remains sustained over time.

Under the above backdrop, an attempt is being made by us to conduct a systematic study of different categories of farmers in order to test whether certain new agricultural practices are undertaken by them on sustainable basis or not.

Specifically, in this study we have made an attempt to test the following hypotheses:

1. That most of the scientific practices adopted by farmers in the study area are still lying at sustainable level.
2. That most of the farmers in the study area have adopted modern agricultural practices.
3. That most of the farmers are aware of the proper doses of application of various farm inputs.

The study has been made on the basis of secondary as well as primary data of agricultural sector of Malda district. Relevant data were collected from various government sources and from field survey. It is recognized that the field survey to be more representative should cover a fairly large number of villages representing sufficiently the entire district. But due to resource and time constraints, we selected only two most agriculturally important blocks, (as believed to be mostly representative) of the district. The two blocks are Chanchal-1 and Bamangola. In Chanchal-1 block we visited Mahandapur, Kharba, Motiharpur, Bhagabatipur, Bhagabanpur, Rani Ka Math and Olihanda villages. But Olihanda and Rani Ka Math seemed to be the representative of the whole Chanchal-1 block for our purposes. So data were collected from these two villages. In Bamangola all the villages appeared to contain the necessary features of the entire block. So Dakatpukur and Parhabinagar villages were selected at random for this study.

From each of the two blocks 30 samples had been taken. Among 30 samples in both cases, 15 samples were taken from each village. Interviewees were the farmers who have at least 10 years experience in cultivation.

This paper is divided into five sections. Section-II makes brief review of literature. Section-III brings out the results of the field survey and makes a discussion of the results. Sections-IV draws conclusion on the basis of analysis of secondary and primary data and finally in section-V we have made some suggestions for the accelerated growth of agricultural productivity keeping in view the sustainable agricultural development.

REVIEW OF LITERATURE

A good number of studies have been undertaken both in India and abroad on sustainable agricultural development. Let us make a brief review of literature on some of the studies undertaken in India.

Monohari (2000) in his study says that the intensive use of inputs has not only polluted the soil, environment and water causing their slow down degradation but also affected the human being and the scientists have realized that the green revolution with high input use has reached a plateau and is now sustained with diminishing returns and falling dividend. After seeing the deleterious effects arising with the use of agro chemical coupled with the degradation of cultivable land and increasing agricultural pollution has created an un- healthy situations in the country, he suggested organic farming. Organic farming aims at cultivating the land and raising crops in such a way as to keep the soil alive and in good health may be an alternative to the present system of farming solely depending upon chemicals. According to him organic farming is a method of farming, which avoids largely, the use of compound chemicals such as chemical fertilizers, pesticides, etc. Instead of those natural resources such as organic matters, minerals and microbes are used. It gives an idea to use all sources, which are natural so that soil health is maintained. He opines that organic farming may be introduced as it relies on large scale application of animals or Farm Yard Manures (FYM), compost crop rotations, cooperative residues green manuring, vermi-compost, bio-fertilizers, bio pesticides and biological control .He thus emphasized that organic agriculture is a viable alternative because it enlivens the soil, strengthens the natural resource base and sustains biological production at levels to commensurate the carrying capacity of the managed agro-system.

In Sangwan's (2005) article we get some pillars to sustainable agriculture. First of all author says that soil health care is fundamental to sustainable agriculture. Ideal soil health care includes provision of optimum soil reactions, soil moisture, content and addition of organic matter. All these factors, according to him, aid in improving soil fertility. Next study is Integrated Water Management. Management of water in agriculture deals with the use of water efficiency under different soil and climatic condition. The author says that agricultural growth and development can be sustainable only if we judiciously use our valuable and limited water resources with the help of modern technologies and traditional wisdom. After water there comes Integrated Nutrient Management. It is a management giving the amount, source, placement, form, and timing of the application of nutrients and soil amendments to ensure adequate soil fertility for plant production and to minimize the potential for environmental degradation, particularly water quality impairment. Again to achieve an evergreen revolution there is a need for Integrated Weed Management so that loss is minimized. He says that another pillar of sustainable agriculture is the use of Remote Sensing in agriculture. It promotes sustainable agriculture through applications of crop health monitoring, crop production forecasting, command area management and so on.

In another article Sangwan (2005) has discussed some philosophies regarding sustainable agriculture. It is said that farmer's involvement is key to sustainable agriculture. Farmers should not get sustainable agriculture in a day, as sudden switch over could spell disaster. To achieve sustainability farmers have to follow two philosophies. One is that anything one does for a small farmer, all farmers' benefit, i.e., for the farmers; the transition to sustainable agriculture requires a series of small and realistic steps. Secondly, for sustainability the tempo of agricultural growth in the country and farm women will have to be given more prominence in agricultural and allied vocations through provisions of critical resources, education and training that will make them full partners in sustainable agriculture.

Dutt and Sundharam (2000) argue that as man by himself can produce a small but with the help of machinery he can produce more, so mechanization is necessary. They also say that farm machinery has relieved man of much of the heavy work as farm machinery helps to perform these tasks easily. Not only that, farm machinery has led to a large-scale production. They cited the example of USA, Canada and Australia, where extensive mechanization of agriculture has taken place agricultural productivity has been raised manifold. They advocated that as India's present problem is to raise production, there is a good case for mechanization of agriculture.

Pandey (1989) says that with the introduction of HYV programme and multiple cropping, the importance of new agricultural strategy has increased manifold with the result that several varieties have been released for cultivation in order to meet the increasing demand of improved seed of dwarf varieties of wheat. According to Pandey in any system of intensive agriculture, the harvesting of crops takes place in succession, often several times a year. This involves a recurring drain of nutrients from the soil and sustained agricultural production at a high level will be impossible unless the nutrient element removed from the soil are regularly returned to it. Nitrogen, Phosphorus and Potash are the nutrients, which are taken by the plants in largest quantity and are removed from the soil during cropping. Therefore, if crop yields are to be increased and maintained, these elements have to be compensated to the soil. He also says that in view of the introduction of HYV the plant protection measure is essential ingredients of the package of recommendation to get desired yield. The application of higher doses of fertilizers promotes vegetative growth and increases vulnerability of these for the attack of pests and diseases.

Sarika (1982) working on the impact of irrigation on small farmers of northeastern region pointed that irrigation project combined with the adoption of improved technology could transform the small farmers into potentially viable units.

Sharma and Sharma (1970), reviewing the problems of small farmers were of the view that by extending the irrigation facilities the problems of small farmers can be solved.

Atteri and Joshi (1982) made a case study in Delhi and reported that besides capital, irrigation plays a crucial role in increasing net return of the farmers.

Sukhpal Singh (2004) in this article studies the factors that led to crisis in Punjab agriculture even after Green Revolution. The factors, he found, responsible for crisis are over capitalization of agricultural sectors; wheat and rice production in Punjab becoming either marginally competitive or non-competitive in international markets ;inefficient use of irrigation; overextension of cultivation of rice led to an increase in insects and pest attacks and last but not least emergence of second hand tractor markets.

In suggestion he quoted the recommendations of Jhol Committee (2000) that 10 lakh hectares of rice and wheat cultivation should be replaced with high value crops like oil-seeds and pulses. By quoting Gill (198) he also says that it is necessary to move farmers with investible surplus from agricultural sector to industrial sector. He himself suggests that the best option is to make farmers partners in agri-business projects initiated by the state and MNCs.

Anwar (2001) in his article analyzed how arsenic is released to ground water. He showed that fertilizers are contaminating ground water in two ways-one is the Pyrite oxidation and the other is Oxyhydride reduction. He says that as there is no standard to arsenic in fertilizer, the department of agriculture should come forward to safeguard public health and make quantitative analysis of fertilizer all over the Bengal basin. In his opinion nitrogenous and phosphate fertilizers are also responsible for groundwater contamination. Heavy rain or irrigation following nitrogen application causes run off and leaching of phosphorus contamination in surface water comes from direct damping of wastes and from eroded suspended solids from urban and agricultural land, he says. He is also of opinion that increased intake of fertilizer is continuously putting additional poisonous ingredients into the soil and causing serious pollution problems and remarkable expansion of chemical fertilizers has caused serious depletion of essential plants nutrients. He suggests that fertilizer invented in temperate climate should not be used in tropical soil as it behaves differently. In this context he also suggests fertilizer management in the crop field, which according to him is crucial issue for maintain the quality of our life supporting resources to meet the demand of the future generation.

Dasgupta (2000) in her article evaluates Green Revolution after four decades of the very revolution. She says that due to green revolution food production registered a phenomenal increase but though agricultural production has continued to increase, the rate of yield per hectare has started to decline. According to her this happens because of over extraction of ground watershed studied various inputs and their effects on Punjab and Hariyana. She found that the growth rate in productivity of rice and wheat has registered a decline in most of the district in Punjab and Hariyana.

RESULTS & DISCUSSION

Since the mid 60's, traditional agricultural practices are gradually being replaced by modern technology and farm practices in India. Initially this technology was confined to only a few districts of Punjab and Hariyana. Later West Bengal was taken into this program. In West Bengal rural agriculture sector got this achievement to some extent later. In our study, we shall consider the case of sustainability of agriculture in relation to certain modern agricultural practices in two blocks of Malda district of West Bengal. The two blocks are Bamongola and Chanchal-1.

USE OF HYV SEEDS

Field survey revealed that majority of the rural people started paddy cultivation with HYV seeds. They mostly use Sarna, Agni, Lalat, Ranjit, Hiramoti, IR-36, IR-64, Taichung, and 40-94, Pari seeds of paddy. These are HYV type of seeds (see Table 1 below:

Table 1
Type of HYV Seeds Used in Two Blocks

Seasons / Blocks	*Chanchal-1*	*Bamongola*
Pre-Kharif (April-July)	IR-64, MTV-7029, UPAS-120	Jala, Prova, MTV-7029, ITE-4094
Kharif	Swarna, Agni. Lalat, Ranjit Hiramoti	Swarna,IR-36,Lalat,Agni BPT-5204,Sumit,Jitendra
(July-Oct)	MTU-7029,IET-14105,Ananda,	
Rabi	IR36, Sarda (Kalai),	Sarda (Kalai), Kasturi Turmaric
(Oct-March)	Kasturi (Turmaric)	
Boro (Jan-June)	Agni, Bipasha	Agni,GolokITE-11898 Sudhir

Source: Field Survey.

These seeds yield more crops to farmers. So they have stopped using Indrashal, Magurshal, Nagra, and Mashuri variety of seeds, as these were traditional seeds. These seeds were mostly used about 10 to12 years ago. Thus with the advent of HYV seeds farmers are now accustomed to use HYV seeds.

Survey report says that out of 60 samples in four villages of two blocks an average of 85% people use HYV seeds. In Chanchal-1 almost 100% people use HYV seeds and in Bamongola about 84% people use HYV seeds. Most interesting fact is that in Chanchal-1 7% people likes to use HYV seeds along with traditional seeds (see Table 2 below)

Table 2
Distribution of Sample Farmers According to Use of HYV Seeds

Name of the blocks	*Whether use HYV seeds*					
	Yes	*Percentage of Total count*	*No*	*Percentage of Total count*	*Mixed Cropping*	*Percentage of Total count*
Chanchal-1	30	100	—	—	7	2.1
Bamongola	28	84	—	—	—	—

Source: Field Survey.

It should be noted that in Bamongola the people who don't use HYV seeds are fully aware of HYV seeds. But they don't practise HYV as because of financial problems.

IRRIGATION AS PART OF NEW TECHNOLOGY

HYV seeds are very thirsty. So during Boro or Pre-kharif cultivation 2-5 times irrigation is most vital. Without proper irrigation application of fertilisers are meaning less. Report says that out of 30 samples in Chanchal-1, 27 people use shallow (both rented and own) machines, tube wells, and pump sets (both rented and own) But nobody use Government source of irrigation as it is very scarcely available in that area. In Bamongola some people use Government source of irrigation. Some others use shallow and others use pump sets (Table 3).

Table 3
Sources of Irrigation Used by the Farmers

Name of the blocks	*Shallow*	*Pump set (Owned)*	*Pump set (Rented)*	*Govt. source*
Chanchal-1	27	1	2	—
Bamangola	14	3	—	—

Source: Field Survey

Table 4
Time of Irrigation Provided to Different Crops (in hour)

Name of the blocks	*Pre-kharif*	*Kharif*	*Rabi*	*Boro*
Chanchal-1	6-8hr	2-5hr	2-5hr	5-10hr
Bamangola	6-10hr	2-5hr	2-5hr	5-10hr

Source: Field Survey.

Thus, it is obvious that no body depends on rain. As a part of green revolution technology people are aware of irrigation work. Farmers of Chanchal provide a total of 6-8 hours of irrigation to pre-kharif crops, 2-5 hours of irrigation to kharif crops, and 5-10 hours of irrigation to Boro crops in a season. Selected farmers of Bamangola provide almost same irrigational facilities to their crops as farmers of Chanchal do (Table 4).

FERTILISER USE

Plants germinated from HYV seeds are dwarf in height but assimilates more plant nutrients from the soil. So, fertilisers play a vital role in supplying nutrients. Farmers of the both Chanchal and Bamangola blocks mostly use Urea; DAP; 10:26:26; Sufola; Super phosphate etc. (Table 5).

Table 5
Main Fertilisers Used in Two Blocks

Crops	*Name of the blocks*	
	Chanchal-1	*Bamongola*
Paddy	Urea, DAP, 10:26:26	Urea, DAP, 10:26:26
Wheat	DAP, Sufola	DAP, Sufola
Mustard	Urea, phosphate	Urea, phosphate
Potatoes	Urea, DAP	Urea, DAP
Pulses	Urea	Urea

Source: Field Survey.

Sometimes they use Borax if plant growth is not satisfactory. It should be noted that they not only use chemical fertilisers but also cow dung and cakes of mustard seed as organic manure.

Investigation reveals that, in both the blocks all large and medium farmers use chemical fertilisers. As far as the marginal farmers are concerned, in Chanchal-1 70% of them use chemical fertilisers and in Bamangola 87% people use chemical fertilisers (Table 6).

Table 6
Distribution of Farmers Using Chemical Fertilisers

Types of farmers	*Name of the blocks*	
	Chanchal	*Bamongola*
Marginal	70	87
Medium	100	100
Large	100	100

Source: Field Survey.

As far as the rate of application of fertilisers are concerned, farmers of Chanchal use upto 60 kgs of nitrogenous fertilisers in a bigha of land, whereas farmers of Bamongola use almost same amount of nitrogenous fertilisers as farmers of Chanchal do. Both the blocks use 20-35 kg of Potassic fertilisers in a bigha of land and they also use almost same amount, viz., 30-40kg of phosphate fertilisers in a bigha of land (Table 7).

Table 7
Rate of Different Types of Fertiliser Applied (Per Bigha)

Types of fertilisers	*Name of the blocks*	
	Chanchal-1	*Bamongola*
N type	40-60	20-60
K type	20-25	20-35
P type	30-40	30-40

Source: Field Survey.

Farmers of both the blocks like to use Borax at the rate of 10kg per bigha if plant growth is not proper.

PESTICIDES, INSECTICIDES AND FUNGICIDES USE

Farmers of both the blocks willingly use pesticides, insecticides, and fungicides. In Chanchal-1 we have 84% and in Bamongola we have 96% farmers who use pesticides. About 70.5% of farmers of Chanchal and 70.8% farmers of Bamongola use insecticides. But the rate of use of fungicides is too much low in both the blocks (Table 8).

Table 8
Distribution of Farmers According to use of Different Plant Protection Measures in two Blocks

Type of plant protection measures	*Chanchal-1*	*% of total % of total block*	*Bamangola*	*% of the total people of the block*
Pesticides	28	84%	29	96%
Insecticides	25	70.5%	26	70%
Fungicides	10	30%	2	0.6%

Source: Field Survey

In both the blocks farmers mostly use Thiodyne, Demirone, Aldrin to protect their crops from the pest and vectors. Both the blocks use same amount of pesticides (Table 9).

Table 9
Amount of Pesticides Applied

Name of pesticides	*Chanchal*	*Bamongola*
Thiodyne	100grms	100grms
Demicrone	100grms	100grms
Aldrin	150ml	150ml

Source: Field Survey.

MECHANIZATION OF AGRICULTURE

People of the study blocks are well exposed to the use of different types of machines in agriculture. In Chachal-1 block about 81% farmers and in Bamangola about 75% of farmers use tractor for the ploughing purpose. In

Bamangola 69% farmers and in Chanchal-1 45% of farmers use power tiller. About 78% of farmers of Chanchal and 87% farmers of Bamangola use pump set to provide irrigation. It should be noted that the figures include both rented and owned. No farmer under the survey area was found to use Harvesters or Threshers (Table 10).

Table 10
Distribution of Farmers Using Machines

Inputs	*Chanchal*	*in % term*	*Bamangola*	*in % term*
Tractor	27	81	25	75
Power tiller	15	45	23	69
Pump set	26	78	29	87
Harvester	—		—	
Thresher	—		—	

Source: Field Survey.

Let us now look at the recommended dose of different inputs, which are mostly used in the two blocks of Malda district.

As soil quality differs from block to block or even from village to village, it is not possible to prepare a chart for the proper doses of different inputs for the different blocks or villages. So experts made a range of dose of different inputs for the farmers and farmers may follow the range and may use different inputs as per requirement within the range.The following table shows the recommended range of irrigation applications.

Table 11
Recommended Dose of Irrigation Per Bigha of Land in Malda District Under Hyv Crops

Low adoption	1 irrigation only
Medium adoption	1-3irrigation
High adoption	4 and above

Source: PAO Office, Malda.

The following table shows the recommended range of fertiliser applications.

Table 12
Recommended Dose of Fertiliser Per Hectare in Malda District Under Hyv Crops

Type	*Low adoption*	*Med. Adoption*	*High Adoption*
Nitrogen	Up to 60 kg	60-80 kg	100 kg and above
Phosphorus	Up to 30 kg	30-45 kg	46 kg
Potash	Up to30 kg	30-45 kg	46 kg and above

Source: PAO Office, Malda.

Here Nitrogen type fertiliser mainly includes Urea. Phosphorous type fertiliser includes Super Phosphate, DAP, Ammonium Phosphate etc. Potash type fertiliser includes Murate of Potash, Sulphate of Potash etc.

Next let us discuss the recommended doses of plant protection measures. Plant protection measures mainly include pesticides, insecticides, and fungicides. There are different brand names for these plant protection measures. The following table shows the rate under broad headings.

Table 13
Recommended Dose of Pesticides, Insecticides and Fungicides in HYV Crops Pre Hectare

Type	*Low adoption*	*Med. Adoption*	*High Adoption*
Pesticides	100-125gm	150-300gm	300gm and above
Insecticides	150-180gm	200-250gm	250-350gm
Fungicides	50-80gm	100-125gm	150-200gm

Source: PAO Office, Malda.

The following table shows the low, medium and high adoption area.

Table 14
Adoption Level Specification

Low adoption area	80% area of cultivation
Medium adoption area	90 % area of cultivation
High adoption area	100% area of cultivation

Source: PAO Office, Malda.

Now if we compare the pattern of use of different inputs we get the following results:

(a) Lower Rate of Chemical Fertiliser Used

In the following table we have compared recommended dose with the dose practised by the farmers.

Table 15
Comparison of Practised Dose of Chemical Fertiliser with Recommended Dose

Types of fertilisers	*Chanchal-1*		*Bamangola*	
	Recommended dose (Kg / bigha)	*Dose followed by farmers (Kg / bigha)*	*Recommended dose (Kg / bigha)*	*Dose followed by farmers (Kg / bigha)*
N type	40-100	40-60	40-100	20-60
P type	30-50	20-25	30-50	20-35
K type	30-50	30-40	30-50	30-40

Source: Field Survey and PAO Office, Malda.

So it is evident from the field survey that farmers of both the blocks are still using lower rate of chemical fertiliser per bigha than the rate recommended by the Agricultural Office of Malda.

(b) Lower Rate of plant Protection Measures used

Like the use of chemical fertiliser, different types of plant protection measures are also being practised at a lower rate. It would be clear from the following table.

Table 16
Comparison of Practised Dose of Plant Protection Measure with Recommended Dose

Types of plant protection measures	*Chanchal-1*		*Bamangola*	
	Recommended dose (Ml / bigha)	*Dose followed by the farmers (Ml / bigha)*	*Recommended dose (Ml / bigha)*	*Dose followed by the farmers (Ml / bigha)*
Pesticides	100-300	100-110	100-300	100-115
Insecticides	150-180	100-110	150-180	100
Fungicides	100-150	40-50	100-150	40-60

Source: Field survey and PAO Offfice, Malda.

So it is also clear from the above table that in both blocks farmers are using plant protection measures at a lower rate. Let us now compare the productivity of some principal crops of Malda district with state or national level. We see the following picture.

Table 17
Comparison of Productivity of Some Major Crops

(Kg //hectare)

Crops	*2000-01*			*2001-02*			*2002-03*		
	D	*S*	*N*	*D*	*S*	*N*	*D*	*S*	*N*
Paddy	1267	1430	1626	1325	1467	1732	1357	1496	1562
Wheat	2511	2336	2708	2616	2400	2761	2592	2612	2618
Maize	1316	1632	1822	1229	1792	1963	1154	1539	1638
Mustard	1083	1012	935	1051	1009	1017	747	877	866
Jute	539	1809	2026	1754	2018	2182	1336	1809	2153

Note: D-district level productivity; S-state level productivity, N-national level productivity.

Source: Annual Action plans of Malda district, Agricultural Hand books of Malda district, Economic Survey (2001, 2002, 2003 issues) and Fertiliser Statistics (2000, 2001, 2002, 2003 issues).

If we now study the productivity of some major crops of Malda district by making a comparison with state level and national level productivity wee see that the productivity of Malda is to some extent behind the state level productivity as well as national level productivity. Though we seldom find a sign of improvement in the production of paddy, wheat, and maize and in jute. But the average rate of yield is not satisfactory as compared to the state level productivity and national level productivity.

CONCLUSIONS

On the basis of the result of the present investigation the following conclusions may be drawn:

Most of the farmers are increasingly adopting new technology in agriculture. The farmers of the district under investigation are greatly influenced by modern technology and on the other hand their dependence on modern technology has been increasing gradually. Most of the farmers had the favourable attitude towards the adoption of new technology. Majority of them are familiar with HYV seeds, irrigation, fertilisers, and plant protection measures etc.

The study shows that farmers using the agricultural inputs at a lower rate than the rate recommended by agricultural scientists. The investigation also reveals that in Malda the use of pesticides and chemical fertilisers, irrigation were becoming a popular practice, still they are being practiced at a lower rate. Thus, there is sustainable use of new agricultural practices when modern inputs like chemical fertilisers, pesticides etc. are being used by different categories of farmers of the district. In future, the use of productive inputs may be increased to the optimum extent to raise the yields of various crops and keep the use of the practices sustainable and hence aaa sustainable development of the agricultural sector.

Lastly, we may conclude that, since modernisation and mechanisation of agriculture through adoption of new technology is only way to stepping up agricultural population, farmers, irrespective of their land size and level of education should avoid traditional agricultural practices and must resort to improve agricultural practices to the extent possible for them.

SUGGESTIONS

On the basis of above analysis the following measures may be suggested for the adoption of new agricultural technology by the farmers keeping sustainable development of agriculture in tact.

1. HYV cultivation for its success depends on fertiliser, irrigation, plant protection measures and other modern implements. Govt should come forward to supply adequate amount of these inputs at a reasonable rate.
2. HYV technology can provide higher yields only when recommended dose of fertilisers, irrigation, and plant protection measures is applied. As the dose applied by the farmers are very low than the recommended rate, the farmers should use upto the recommended rate of these inputs. Efficacious and judicious use of fertilisers should be initiated through extension workers and various media for the farmers.
3. As per the study, Govt. source of irrigation is very low and it is shallow-tube well, pump set irrigation that is increasingly being used by the farmers of all categories as source of irrigation. But small and marginal farmers can hardly afford to purchase costly pump sets due to their poor economic condition. So provision should be made to supply pump

set at a subsidized rate. However, farmers should refrain from wasting irrigation water.

4. In most cases supply of electricity to tube-well and pump sets is in adequate and irregular. Most of the farmers using pump set have to use diesel as fuel for running their pump sets. But due to the high price of diesel many farmers can't afford to purchase the required amount of diesel for fuelling their pump sets. It may be suggested that a differential price system favouring the farmers should be introduced.
5. In addition to the use of chemical fertilisers farmers should be encouraged to use organic and compost fertiliser. Organic and compost manure can retain the original fertility and productivity of land and hence sustainability of agriculture.
6. The farmers are not using modern inputs at a recommended rate because of lack of money. So credit facilities through banks and co-operative institutions should be made available to the farmers. Not only that, a procedure of need based and asset free credit facilities should be introduced on preferential basis.
7. Adoption and efficient utilisation of technology depends on the knowledge and education of farmers. Majority of farmers are still ignorant and illiterate and to a considerable extant they are governed by traditional belief. Govt. should provide more education facilities in the viliages where farmers live.
8. Demonstration of improved practices should be made broad based and for that purpose small farmers are to be selected rather than larger farmers. Village level trained extension workers should be made more responsible for carrying out more practical works in these demonstrations.

REFERENCES

Abrams, Albert (1916), New Concepts in Diagnosis and Treatment. Philopolis Press, San Francisco, CA.

Alvares, C. (ed.) (1996), *The Organic Farming Sourcebook*. The Other India Press and Third World Network, Mapusa, Goa.

Amin, K., Kothari, A. and Rao, S. (1991), Whither Ecological Farming? Kalpavriksh, New Delhi (unpublished report).

Anwar, J., "Arsenic and Uranium in Fertiliser", *www.sos.arsenic.net/english/tsp.html.*

Banwari (1992), *Pancavati: Indian Approach to Environment*. Shri Vinayaka Publications, New Delhi.

Barr, Sir James (ed.) (1925), Abram's Methods of Diagnosis and Treatment. W. Henemann, London, U.K.

Bhatia, S. and Kothari, A. (1996), Community Register for Documenting Local Community Uses of Biological Diversity. *Bulletin of the Working Group on Traditional Resource Rights* No. 2, Spring.

Chandrasekhar, P. and M.S. Nagraja (1996), "Utility of Rock Phopspate in Sustainable Agriculture", *Yojana*, October.

Chaudhary, T.N. (1997), Water Management In Rice For Efficient Production. Directorate of Water Management Research, Patna, India.

Dasgupta, Kumkum, "The Green Revolution in India–Is the Joyride Over?", *www.mindfully.org/farm/IndianGreenRevolution.*

Denning, R. Murray (1988), My Search for Radionic Truths. Borderland Sciences Research Foundation, Garberville, C.A. 118 p.

Dharampal (1983), *Indian Science and Technology in the 18th Century*. Academy of Gandhian Studies, Hyderabad.

Gadgil, M. (1996), People's Biodiversity Register: A Record of India's Wealth, Amruth, Vol. 1 No. 5, October.

Gupta, A. (1996), *Roots of Creativity and Innovation in Indian Society: A Honey Bee Perspect.ve*. Lovraj Kumar Memorial Lecture, Society for Promotion of Wastelands Development, 30 August, 1996, New Delhi.

Gupta, P.K. (1986), *Pesticides in the Indian Environment*. Interprint, New Delhi.

Gupta, R.K. and I.P. Abrol (1990), Salt Affected Soils their Reclamation and Management for Crop Production. Adv. Soil Sci. 11: 223-288.

Gupta, R.K., N.T. Singh, and M. Sethi (1994), Ground Water Quality for Irrigation in India. CSSRI Technical Bulletin 19. Central Soil Salinity Research Institute, Karnal, India. CSSRI, Karnal, 1994. p. 16.

Indian Society for Rural Gene Banks (1992), *Rapid Clonal Multiplication of Rice Seed: A Field Guide*. The Other India Press, Goa.

Kothari, A with Kothari, M. (1995), *SacrificingOour Future: The New Economic Policy and the Environment*. Consumer Unity and Trust Society, Calcutta.

Kothari, A. (1999), Intellectual Property Rights And Biodiversity: Are India's Proposed Biodiversity Act And Plant Varieties Act Compatible? Paper presented at Workshop on Biodiversity Conservation and Intellectual Property Rights, 29-31 January, 1999, Research and Information System, Kalpavriksh, and IUCN-The World Conservation Union, New Delhi.

Kuepper, George (1991), Psychotronics: A New Frontier In Sustainable Agriculture. *Natural Food And Farming*. May/June, p. 13, 14, 33.

Monohari, "Organic Farming for Sustainable Agriculture", *www.intrac.org/docs.php2. / www.manage.gov.in/faculty/monohari1.hlin.*

Nijar, G.S. (1996), *In Defence of Biodiversity and Indigenous Knowledge: A Conceptual Framework and the Essential Elements of a Rights Regime*. Third World Network, Penang.

Padmanabhan, N.R., "Reasons for Use, Problem and Market for NEEM Pesticides in Tamil Nadu", *www.pseb.org.pk/studies/Bioinformatics%20ReportDraft.*

Prescott-Allen, R. & Prescott-Allen, C. (1983), *Genes from the Wild*. Earthscan, London.

PVFRPB (1998), Plant Varieties and Farmers Rights Protection Bill, 1998. Ministry of Agriculture, Government of India, No. 18-136/97/SD-IV.

Rao, V.K.R.V., "New C hallenges Before Indian Agriculture" (Dr. Pause Memorial Lecture).

Razzaque, M.A., M. Badaruddin, and C.A. Meisner (ed.) (1995), Sustainability of rice-Wheat Systems in Bangladesh. Proceedings of the National Workshop, Dhaka, Bangladesh, 14-15 Nov. 1994. Bangladesh and Australia Wheat Research Centre, Dhaka, Bangladesh.

Richharia, R.H. and Govindaswamy, S. (1990), *Rices of India*. Academy of Development Sciences, Karjat, Maharashtra.

Sahgal, J.L., Mandal, D.K., Mandal, C. and Vedivelu, S. (1992), *Agro-ecologicalRegions of India*. Technical Bulletin. National Bureau of Soil Survey and Land Use Planning, Indian Council of Agricultural Research, New Delhi, and Oxfordand IBH Pub. Co.

Sharma, P.K. and B. Mishra (2001), "Effect of Burning Rice and Wheat Crop Residues: Loss of N, P, K and S from Soil and Changes in the Nutrient Availability". *Journal of the Indian Society of Soil Science*. 49:425-429.

Shiva, V. (1991), *The Violence of the Green Revolution*. Third World Network, Penang.

Shiva, V., Ramprasad, V. and Holla-Bhar, R. (1994), *Sustaining Diversity: Renewing Diversity and Balance Through Conservation*. Navdanya, New Delhi.

Shiva, V., Ramprasad, V., Hegde, P., Krishnan, O. and Holla-Bhar, R. (1995), *The Seed Keepers*. Navdanya, New Delhi.

29

APPROACHES TO SUSTAINABLE DEVELOPMENT OF RURAL POOR THROUGH SELF-HELP GROUPS

K.P. Kumaran

INTRODUCTION

In spite of massive financial assistance of credit and subsidy provided under various poverty alleviation programmes, the achievement made in terms of household crossing poverty line income levels had not been up to our expectation. Recent studies conducted by experts to assess poverty rate showed that it varied from 20 to 25 percentages in the rural areas. One of the factors responsible for the prevalence of poverty in the rural areas is attributed to poor implementation of development programme and lack of participation of the targeted poor in designing and implementing developmental programmes meant for them. The fact is that no serious efforts were being made to consult the beneficiaries about their felt need either while preparing or implementing the programmes. In other words poverty alleviation programmes were generally thrust up on the beneficiaries. More over, such programmes generally characterized by inadequate credit coupled with lack of infrastructure and expertise, often proved either unproductive or some times as a burden to the beneficiaries. Such a situation often led to" dependency syndrome" among the poor. There fore, there is a need to make developmental programmes need based and the same may be planned and implemented with the participation of the beneficiaries.

In India, participatory development is not new. It is almost synonymous with India's planned development. The first major attempt made by the government to achieve peoples participation was by promoting a large number of co operative and multi co operative societies to provide credit and marketing facilities to farmers. Setting up of Panchayat Raj institutions on the basis of Balwant Rai Mehta Committee in 1972 formed the other attempts on these lines. In addition to this, to promote participatory development among the poor, efforts were made by promoting and sponsoring voluntary organization. However, all these attempts could not succeed in ensuring sincere participation particularly that of poor, either due to rigidity of rules and regulations, domination of its activities by local elite's, politicians or dominant caste members or lack of conducive working environment (Jain: 1985). What explicit in the above attempt is that formal and heterogeneous organizations need not necessarily attract the poor to enlist their participation in developmental activities. What they need is viable informal and homogenous people's organization. In response to this efforts are being made to promote people's organizations at grass-root level, and this resulted in the proliferation of Self-Help Groups (SHGs) among the poor. (Fernandes: 1987).

In this paper an attempt is made to examine the important characteristics of SHGs of the poor and the various approaches that are followed to promote sustainable development among the rural poor. Before we examine these aspects in details let us see the concept of sustainable development.

SUSTAINABLE DEVELOPMENT

The Human Development Report 1991 of the United Nations Development Programme defined sustainable development as "development that improves health care, education and social well being". As sustainable development is achieved through participation it is also defined as organizing the local community into grass root level organizations to enable them to participate in the multi functional task of formulating and implementing projects in areas of income generation, health, drinking water, education, poverty etc., (Hussain: 1991). However, the concept of sustainable development may also vary depending on the degree of development achieved by the local community. In India, given the low social-economic background of the rural poor, the emphasis ought to be on capacity building through awareness generation and training. In addition they should have adequate financial and technical support to undertake viable income generating activities. Therefore, in this paper sustainable development is defined as a process of organizing the poor into grass root level organizations and enhancing the capacity of its members to enable them to achieve socio-economic development.

CONCEPT AND CHARACTERISTICS OF SHGs

This paper deals with informal SHGs of the poor. SHG is defined as a special form of voluntary organization generally formed by a small homogeneous group of people to attain certain goals either social, economic or both. (Kumaran: 1997). They are to a large extent self-supporting and self-governing with limited interventions from outside agencies. Its risks, costs and benefits are shared among its members (Verhagan 1997). In this sense SHGs are people's organization, which give ample opportunities to its members to take part in all activities of the group including decision-making. Another important feature of this group is that they evolve and function on the basis of felt needs of the people largely making use of local resources (Sukhpal Singh, 1995). The size of the SHG generally varies depending upon the purpose for which the group is formed. What is important in this regard, that the size may be compact enough or such that all the members get equal opportunity to participate in all the activities of the group including decision making. This is the main feature that distinguishes SHGs from other groups (Singh Kartar and T.S. Raji Gain: 1995).

FORMATION OF SHGs

One of the reasons for growing poverty in India is attributed to lack of organizing capacity of the poor to organize themselves in to groups to fight against exploitation and poverty. Therefore, for the removal of poverty, the poor should organize themselves into groups to enable them to articulate their rights and demands collectively and participate in their own developmental programmes. However, the fact is that the poor do not organize themselves to fight against their common problem including poverty. Several impediments with in and outside the community resist them from organizing into such groups. The factors which affect within the community include their powerlessness, lack of unity, and leadership qualities and the factors that act from outside include caste, class, and ethnic stratification (Saxena: 1991). Of late, it has been explored that SHG can serve as a good platform for the poor to collectively fight against poverty and related problems faced by them through a participatory mode. In India SHGs are generally formed among women. Their low literacy rate and socio-economic background itself stress the need for an external agency to mobilize and promote group among them. Therefore, except for those formed spontaneously, most of the SHGs are promoted by external agencies such as government or non-government.

SHG PROMOTION INSTITUTIONS (SHPI)

In India, SHGs are generally promoted by non-governmental organizations. There are a good no. of well established NGO engaged in promoting SHGs. Among them, the Mysore Resettlement and Development Agency (Myrada) based in Bangalore, is pioneer and most successful in promoting and sustaining SHGs mainly among weaker sections. Over the years this organization promoted several groups among the poor particularly in the southern states of India. Others include Professional Assistance for Development Action (PRADAN) and Association of Sarva Seva Farms (ASSEFA) in Tamilnadu, Society for Helping and Awaking Rural Poor Through Education (SHARE) and Rashtriya Seva Samithi in A.P. Self Employment Women's Association (SEWA) in Gujarath and Sewa Mandir in Rajasthan to name a few among the accredited NGOs.

At the Government level, the Ministry of Rural development takes great interest in promoting and sustaining SHG among the poor. The recently introduced Swornajayanti Gram Swarajgar yojana (SGSY) is in this direction. Under this scheme, self-employment programme is implemented through group approach by making provision for adequate credit, infrastructure, market support and capacity building to take up the venture. Of late several commercial banks, regional rural banks including co-operative banks are actively involved in promoting Self-Help Groups among the poor. Thus from the above explanation it is clear that both governmental and non-governmental agencies involved in promoting and sustaining SHGs among the poor.

More or less similar strategies are adopted by both the agencies while forming the group. In general, trained and experienced grass root level workers through the process of social mobilization form the groups. Several precautions and care are taken by these agencies while forming the groups to retain its special features like viability, homogeneity and cohesiveness so as to make it participatory and self-sustaining.

APPROACHES TO SUSTAINABLE DEVELOPMENT

Earlier we have seen that SHGs are generally formed to meet certain felt needs of people. Based on these needs the socio-economic activities undertaken by them vary from group to group. To sustain these developmental activities, several strategies such capacity building of SHGs so also the promotion institutions, promotion of thrift and credit activities, income generation activities and linkage with micro-finance and panchayati raj institutions are undertaken. Let us examine each one of them in detail.

CAPACITY BUILDING

One of the approaches followed for sustainable development of SHGs is capacity building of its members. This is very important as most of the group members are illiterate and from poor socio-economic background. Therefore, for the effective management and administration of group activities, its member's capabilities are to be scaled up. This is attempted through a two pronged strategy viz., awareness creation and training. Awareness creation is achieved through informal education to sensitize about the problems and make them aware of their rights and responsibilities. The ultimate objective of awareness building is intended towards achieving social development.

Training is another important component of capacity building. Through training, knowledge and skills are imparted to its members. Training helps the members to upgrade their skills and knowledge to undertake various income generation activities. It also equips them for the overall administration and management of the SHGs. Experience in the Indian context showed that those SHGs undergone capacity enhancement through proper training to a large extent successful in promoting sustainable development among its members (Kumaran: 1997).

The best example where systematic approach is followed in enhancing capacity of SHG and its members for poverty alleviation of rural poor through rural credit is that of International Fund for Agricultural

Development (IFAD) assisted project in Tamilnadu (Tamilnadu womens Development Project) and Maharashtra (Maharashtra Rural Credit Project). In both the states the respective projects were implemented in partnership with NGOs and the Government. As mentioned earlier under the project capacity building of SHG members is an important component jointly undertaken by NGOs and specialized agencies. For eg. In the latter, Maharashtra Industrial and Technical Consultant organization (MITCON) and Maharashtra Centre for Entrepreneurship Development (MCED) take active role in promoting the capacity building of SHG entrepreneurs. In both the cases adequate funds were earmarked for capacity building of SHG members and the entrepreneurs. (See Kumaran: 1997) Similarly, some of the banks and Govt. department, which are actively involved in promoting SHGs, have their own resources and manpower including infrastructure for such activities. However, the reality is that most of the SHGs are promoted by NGOs, which do not have these facilities and resources for capacity building. Nevertheless, there are few sources from where they can avail financial assistance for such activities.

Currently National Bank for Agriculture and Rural Development (NABARD) is providing promotional grant to NGO for various purposes like meeting the cost of promotion of SHG and training infrastructure. But the number NGOs benefited from those programme is very limited. Similarly, the margin money received by NGOs through interest from Micro-finance institutions like NABARD, Rashtriya Mahila Kosh (RMK) and Small Scale Industrial Development Bank of India (SIDBI) is not very attractive and sufficient . Most of the NGOs participating in this programme mainly meet these expenses from the funds or grants received from donor agencies. While those NGOs who do not have such source of grants are to be provided with technical and financial assistance to enable them to take up capacity building of the SHGs promoted by them.

THRIFT AND CREDIT

Lack of availability of timely institutional credit at reasonable interest is considered as one of the factors for increasing poverty among the poor. Therefore, the poor mainly depend on informal agencies including moneylenders in meeting their credit needs at a higher rate of interest. Inability to repay the loan on time results in the accumulation through compound interest making them unable to come out of the debt trap. One of the Strategies adopted by the SHGs to reduce dependency on credit from outside agencies is by mobilizing financial resources available with the group members through thrift and credit. And this is the entry point for any SHG. Mobilizing internal resources is important to ensure the sustainability of the group even after the withdrawal of external funds. Depending on the ability, the group members save small amount of money ranging from Rs. 10 to 50/- on weekly or monthly basis and the saved amount is pooled together and use it for internal lending at an interest rate ranging from 2% to 5% per month among its members.

The quantum of amount to be saved, the interest rate of internal lending, number and amount of loan to be disbursed including identification of lonees are determined by the members themselves. Usually the repayment of loan is around 95 per cent and if it is less not acceptable to the group. In SHGs usually the internal lending start after 6 months of its existences and strict discipline is maintained through peer group pressure and other means for timely repayment of loans. Always transparency is maintained in all financial and other transactions by ensuring participation of all members in most of the activities.

The pooled saving is lent out among the members for meeting their emergent needs mainly for consumption. In some cases, the amount is also used for undertaking small economic activities depending on the availability of funds. Savings of the SHGs particularly those promoted by certain NGOs run into several lakhs and even in crores. In such cases Women's Bank were established on the line of celebrated Women's Bank in Bangladesh. Examples are Rural Women's Bank in Madurai. Tamil Nadu and SEWA Bank in Gujarath, etc.

With a view to make thrift and credit activities meaningful for economic empowerment, the groups are generally encouraged to invest the saving for income generation activities. In a survey conducted by NABARD (1989) identified several SHGs engaged in a variety of non-farm activities ranging from fruit preservation, savary making, bamboo works, leather works, beedi making and trading of fruits etc in the state of Kerala, Tamil Nadu, Rajathan, Madhya Pradesh and Maharashtra.

In another study undertaken by Sukhpal Singh (1994) it is found that apart from non-farm activities SHGs are successful in undertaking agri-business particularly in the states of Bihar, Punjab and Gujarat. The group ventures undertaken through SHGs are best strategy for the socio-economic empowerment of the rural poor. (NABARD: 1989; Raji, Gain, 1997).

LINKING SHGS WITH MICRO FINANCE INSTITUTIONS

As mentioned earlier, the money mobilized internally is pooled together and generally used for meeting emergent consumption needs. If surplus is available used it for undertaking economic activities. In such cases the internally mobilized resources wont be sufficient for undertaking income generating activities. There fore tapping of external credit is very important. With the objective of providing supplementary credit to the poor in cost effective and sustainable manner in 1992, NABARD launched pilot project for linking SHGs with Banks. There are 4 distinct models of linking between SHGs and banks. The simplest and most direct is a model in which the bank deals directly with the individual SHG members. Another model is where the bank gives direct assistance to the SHG and SHPI provides training and guidance to the SHG. The third model involves the SHPI as a financial intermediary between the bank and a number of SHGs. Here the SHPIs take up responsibilities for repayment to the bank. The fourth model envisages bank loans directly to individual members of SHG up on the recommendation of the SHGs and the NGO. In this case the NGO assists the bank in monitoring supervision and recovery of loans (Nanda: 1995).

The pilot project of linking of SHG with banks started in the year 1992. Ever since the NABARD started the bank linkage programme the progress has been slow but steady. During 1992- 1999 the number of groups financed was 32,995 and the amount disbursed amounted to Rs. 571 million. By the end of 2003-04 the number of SHGs linked went up to 1.1 million and amount disbursed stood at Rs. 39,042 million. (NABARD: 2003-04). Delivery of credit under this programme is disbursed through nationalized banks, Co- operative banks and NGOs. The above data showed that initially the coverage was very slow and in due course it picked up. But over the years the extent of coverage achieved in terms of physical and financial terms is not very encouraging. Apart from NABARD, RMK and SIDBI also provide micro credit to NGOs for on lending to SHGs. Here also the coverage is too small By March 1998, the RMK has disbursed around US$ 5.7 million to NGOs and the SIDBI US $2.8 million to 79 NGOs for on lending to SHGs (Paul Mc Guire, 1998). In the years that followed there is substantial improvement in the situation of credit disbursement.

Considering the large no. of SHGs promoted by the NGOs and the vast number of members living below poverty line, the no. of SHG linked and the extent of credit made available to its members through linkage is not at all satisfactory. Only those NGOs, which are well established with dynamic leadership, are able to link the SHGs with bank. Others either due to ignorance about the linkage programme, apathetic or indifference attitude of bank managers or due to lack of motivation on the part of SHPI or SHG members does not come forward to avail the benefit. Although peer group pressure is considered as one of the reasons for better repayment of loan but it should not act as a limiting factor in motivating its member to have access to bank loans.

In a recent study conducted among the women members of the Grameen Bank in Bangladesh it was reported that to ensure timely repayment of the loan the borrower peers inflict intense pressure on women clients. In order to maintain regular repayment schedule through a process of loan recycling considerably increased the debt lending to frustration and tension among household members (Aminur Rehman: 1999). Such a situation emerged as a result of violation and dilution of the basic principles of self-help. The members should avoid such incidence by emulating the spirit of helping those who are in distress.

Apart from linking the SHGs with financial institutions limited attempts were made by some of the promoting agencies to link them with some of the Government sponsored welfare and Poverty Alleviation Programmes. In recent years India has embarked upon a major process of decentralization of its administration through Panchayati Raj institutions to enable the poor to participate in nation building. Therefore it is desirable to link SHG with panchayati raj institutions for participatory development and convergence of welfare and poverty alleviation programmes for their benefit.

LINKING WITH PANCHAYATI RAJ INSTITUTIONS

The 73rd Constitutional Amendment Act enacted in 1994 through the process of decentralization to ensure the rural poor to participate in the developmental programmes that matters them. Although gram sabha has given statutory powers to serve as a major platform for the poor to participate in the developmental activities and express their desires and demands, it does not have any provision to ensure the participation of poor in planning and implementing developmental programme meant for them. In such a situation it is desirable to link SHGs with the panchayati raj institutions. This will not only facilitate the poor to plan and implement developmental programmes based on their felt needs but also identification of beneficiaries among themselves under various welfare programmes of the government. Such an attempt definitely leads to integrated development of the poor by ensuring the benefit of development to the poorest of the poor. However, such an attempt should ensure limited interference from outside agencies and should net distort the autonomy of SHGs. Also important is allow the SHGs to retain all their distinct characteristics.

In this direction, only limited attempts were made by promoting agencies particularly Government in linking SHGs with panchayati raj. In such cases these groups could converge some of the Govt. sponsored welfare and poverty alleviation programmmes for the betterment of the poor. A host of similar other programmes are also available for the welfare of the poor. Efforts should be made to converge these programmes through SHGs for the benefit of its members.

CONCLUSIONS

In India, the development strategy directed towards weaker sections particularly the poor has been undergoing change. Till 1970's the focus was on welfare-oriented programmes and subsequently shifted to participatory development. The need for such a shift necessitated to ensure sustainable development of poor and to enable them to come out of the vicious circle of poverty. In this paper we have examined various approaches followed by SHG for sustainable development of rural poor. The initiatives followed included capacity building of group, promotion of thrift and credit programmes and linking the group with financial and panchayati raj institutions.

Different approaches have been followed by the promoting agencies for the capacity enhancement of the groups. We have seen that the SHGs are largely promoted by NGOs. It is true that most of them do not have adequate infrastructure, including financial resources to take up capacity building programme. Some of the NGOs meet this expenditure from the grants received from funding agencies. The promotional grant provided by the micro finance institutions towards this end is very meager. Therefore, those NGOs whch do not have such resources, there is a need to provide technical and financial support to equip them to perform effective role of promotion institutions.

Thrift and credit activity form another important strategy adopted by the SHG to reduce dependency on external agencies for financial resources. The amount saved is pooled together and used for internal lending among the members to meet their consumption and production needs. However, it is found that the internal saving not sufficient enough to embark on economic activities including income generation activities. In such cases linkages are also established with micro finance institutions for credit at a lower rate of interest.

In India, under the control of Government there are banking and non-baking financial institutions with the commitment of extending micro finance to the SHGs at a reasonable rate of interest. Notable among them are NABARD, SIDBI and RMK. Despite the role played by these three agencies the extent of coverage in physical and financial terms is not very satisfactory. It is found that only those well established NGOs with dynamic leadership were successful in establishing the linkage, while others failed either due to ignorance on the part of SHPI or lack of interest shown by the members. Such a situation demands wider publicity about the linkage programme and need for motivation among the members. Similarly, no serious attempt has been made in linking SHG with panchayati raj and it is still in the initial stages. Therefore there is a need to make maximum linkage of SHGs with panchayati raj institutions mainly to converge welfare and the poverty alleviation programmes for the benefit of poor for their integrated and sustainable development. However, it is cautioned that such linkages should not distort the autonomy of SHGs.

In this paper it is argued that SHG is a best platform for the poor to come together and work together for their sustainable development. Currently, SHG activities are restricted to certain pockets and region of the country. There fore, there is a need for both the Government and non-Governmental agencies to promote such groups in all the parts of the country particularly where the poor are concentrated and make it a national movement.

REFERENCES

Ajay Jain (1998), Poverty Alleviation through Social Mobilisation, the Kurnool Strategy DRDA Kurnool, Andhra Pradesh.

Aminur Rahman (1999), Micro-Credit Initiative for Equitable and Sustainable Development: Who Pay? World Development Vol. 27, No. 1.

Akumal Hussain (1991), Economic Deprivation, Erosion of the resource base and the Imperative for Sustainable Development in South Asia in Participatory Development Learning from South Asia, Co-authored by Punna Vignaraja & Others, Oxford University Press.

Fernandes P., Aloysius (1987), NGOs in South Asia People's Participation and Parnership; World Development Vol. 15, Supplement.

Jain L.C. (1985), Grass without Roots: Rural Development under Government Auspices, Sage New Delhi.

K.P. Kumaran (1997), Self-Help Groups: An Alternative to Instiutional Credit to the Poor: A Case Study in Andhra Pradesh, Journal of Rural development Vol. 16, (3).

Kumaran K.P. (1997), Country paper on Capacity Building of Grass Root Level Organisation in India, APO, Unpublished.

NABARD (1983), Studies on Self-Help Groups of the Rural Poor, Bombay India.

NABARD, 2003-04, Progress of SHG Bank Linkage, Bombay.

NABARD (1994), MRCP: A Sage of People's Participation; NABARD, Pune.

Y.C. Nanda (1996), Significance of Establishing Linkage of SHGs with Banks; Papers Presented in the International Seminar on Development of Rural Poor Through SHGs; NABARD, Bombay.

Paul B. Mc. Guire (1998), Second tier Micro-finance Institutions in Asia in Small Enterprises Development Vol. 9 No. 2.

T.S. Raji Gain (1997), Empowering Women Through Group Enterprises Asian Entrepreneur; Vol. 6:1, January–April.

Saxena K.B. (1992), Rural Organisations: Occasional Papers–3 Govt. of India, Ministry of Rural Development, Krishi Bhavan, and New Delhi.

Singh Katar and T.S. Raji Gain (1995), Evolution and Survival of SHGs; Some Theoretical Proposition, and Empirical Evidences Reflection from Case Studies, BIRD–Lucknow.

Singh, Sukhpal (1995), Self-Help Groups in Indian Agri-business; Reflections from Case Studies.

Srinivasan, Girija (ND), Group Approach to Empowerment of Rural Poor Women –IFAD Experience in Tamil Nadu State, BIRD, Lucknow.

Verhagen Koenraad (1987), Self-Help Promotion: A Challenge to the NGO Community CEBEMO Royal Tropical Institute of Nether Lands.

30

RURAL NON-FARM EMPLOYMENT, RURAL INFRASTRUCTURE AND MODERN FARM PRACTICES: THEIR EFFECT ON AGRICULTURAL PRODUCTIVITY

Rudra Prakash Pradhan

INTRODUCTION

Agriculture continues to be kingpin to all strategies of planned socio economic development in any developing countries in the world. In India, it occupies the most strategic position and views as a 'way of life'. It is one of the largest private enterprises in the country and continues to dominant the change of economic set up through its casual links with factor and product markets (Radhakrishna, 2002). Agriculture is a pre-requisite in India, although industrialization has come to stay as a synonym of economic growth. Its importance significantly lies on rural employment and rural livelihood (Hashim 2004). While it provides livelihood to over 60% of population, it employs about 60% of total workforce of the country (Ranganathan, 2003). The importance of agriculture can also be judged through other aspects like contribution towards GDP, alleviation of rural poverty, empowering agrarian folk, contribution towards export, promotion of economic growth (Sen, 2003; Arnon, 2001; Datt and Raviallion, 1998; Ahluwalia, 1978) and an absolute promoter of rural development (Singh, 2002).

But in the recent time, the performance of agriculture is not satisfactory in the Indian economy. There is declining share of agriculture towards GDP, falling growth of agriculture, lack of diversification and slow down of employment absorption capacity, etc. (Pradhan, 2005). Considering a particular instance, it has observed that while the growth rate of agriculture is about less than 2% per annum in the last 10 years (Dev, 2006), employment growth in agriculture has been substantially negative during 1994-2000 (See Figure 1). Though there are many factors involved in this issue, but we presumed that it is mostly due to lack of coordination with rural non-farm sector, low availability of rural infrastructure, and slow modernization of agriculture. Accordingly, present paper seeks to examine

Figure 1
India's Employment Growth during 1980s and 199

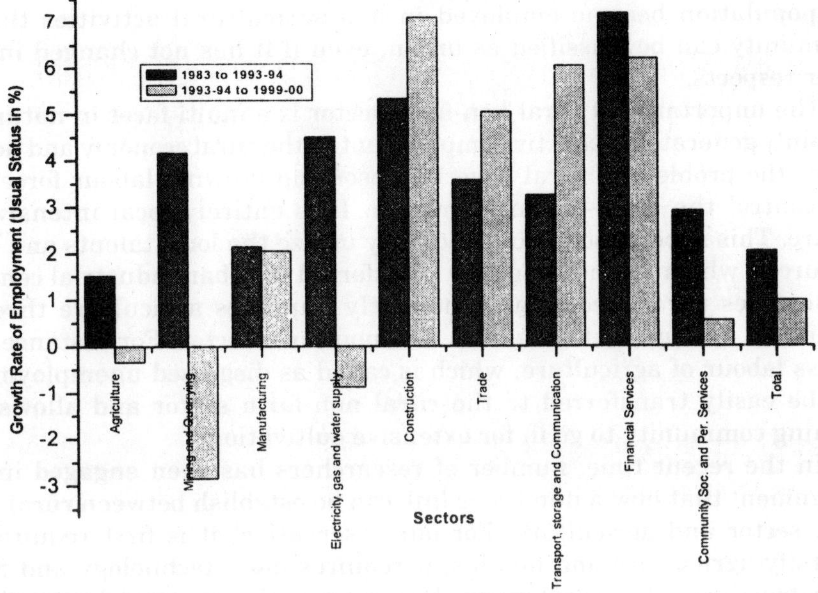

Source: Planning Commission, Government of India.

the impact of rural non-farm employment, rural infrastructure and agricultural modernization on agricultural productivity in the Indian economy.

The rest of the paper is organized as follows. Section II outlines an overview of rural non-farm economy. Section III describes the concept of rural infrastructure. Section IV reveals the importance of modern farm practices. Section V describes the analytical framework and its results are discussed in section VI. And the final section follows conclusion and suggestions.

RURAL NON-FARM SECTOR AND ITS IMPORTANCE

Rural non-farm sector generally defines as a wide spectrum of activities in the rural areas, which are not directly associated with agriculture and its allied but represents very important part in the rural economy in terms of income and employment. The judgments about the viability and importance of the rural non-farm sector hinges crucially on what is meant by 'rural'. Many countries apply different definitions when designating areas as rural or urban. But usually the distinction between rural and urban employment is based on the place of residence of workers so that those who commute to a job in a nearby urban center are considered to be rural workers. Rural is most often defined to include settlements of about 5000 or fewer inhabitants. It is also based on population size and/or functions and characteristics of the settlement like school, hospital, seats of local government, etc. In India, whether or not a particular setting is designated as rural or urban is entirely based on the composition of workforce in that setting. According to Lanjouw and Feder (2001), when more and more of the population become employed in non-agricultural activities than a community can be classified as urban, even if it has not changed in any other respects.

The importance of rural non-farm sector is a multi-facet in nature. It certainly generates productive employment in the rural economy and hence, solves the problem of rural poverty, absorbing growing labour force and can control the rural-urban migration. It is entirely local intensive in nature. This is because it can effectively utilize the local talents and local resources, which can not be easily transferred to urban industrial centres. It increases rural income and indirectly improves agriculture through numerous linkages between farm and non-farm sector. For instance, the excess labour of agriculture, which is called as disguised unemployment, can be easily transferred to the rural non-farm sector and allows the farming community to go in for extensive cultivation.

In the recent time, number of researchers has been engaged in the assignment that how a numerous link can be establish between rural non-farm sector and agriculture. For our observation, it is first required to diversify agriculture and for this, it requires more technology and more infrastructure in the rural areas. As you know the unutilized potential of food-processing in India is enormous. An expansion of this sector is an ultimate way to bring industry into rural areas, expanding the value chain of agricultural production, providing assured markets for farmers, enabling them to diversify into higher value horticultural crops and can expand employment by creating high quality non-agricultural work opportunities in rural areas. Though there can not be a one policy package for the entire rural non-farm economy, a sub-sectoral policies in different regions are to be considered necessary. One of such possible areas is to expand the availability of rural infrastructure and adoption of modern farm practices, which can develop rural non-farm sector and can establish the linkage between farm and non-farm sector.

RURAL-INFRASTRUCTURE AND THEIR IMPORTANCE

Rural infrastructure is commonly defined as a physical framework of facilities through which goods and services are provided to the public. To put it in other way, it is the basic system and services in the economy without which no economic activity can be undertaken in the production process. It provides facilitative set-up that stimulates the development of an economy and is considered as a serious constraint on country's growth potential (GOI, 1999-2000). It's true that infrastructure's installations do not produce goods and services directly but builds Social Overhead Capital (SOC) for all other economic activities. It raises productivity of other factors and hence, ensures more production and facilitates economic growth. For this reason, infrastructure is often called as 'unpaid factor of production', as its availability leads to higher returns obtainable from other factors of production. The pay-off from better infrastructure services goes beyond reducing technical inefficiencies and financial losses. Infrastructure is directly related to other economic activities, since the expansion of this sector leads to expansion of other sectors by meeting demands for it and vice-versa. It is, however, important that when an economy develops, infrastructure must adapt itself to support the changing patterns of demand. Consequently, an adequate quantity, quality and reliability of infrastructure are important pre-conditions for overall economic growth and development (WDR, 1994).

Infrastructure particularly rural one is said to strengthen the foundation of agriculture. It helps agriculture by increasing production and/or productivity and draws excess labour force from agriculture, which results the raise of income of poor in other sectors (Ravallion, 1991) and reduces the incidence of poverty (Ut et al., 2000). But it is important that rural infrastructure is generally viewed as essential requirements for growth of the rural non-farm economy. An inadequate and poor quality of infrastructure imposes serious costs on virtually all-economic activity in the rural areas. The National Policy on Agriculture (GOI, 2001) has emphasized that the availability of rural infrastructure supports agriculture at a much faster rate, promote value addition, accelerate the growth of agro-business, create employment in the rural areas, secure a fair standard of living for the farmers and agricultural workers and their families, discourage migration to urban areas and face challenges arising out of economic liberalization and globalization. The expansion of productive infrastructure services into rural areas is one of the obvious policy recommendations for promoting the rural non-farm sector. There are innumerable studies, which document the limited access of rural enterprises to power and telecommunications, and the high transactions costs caused by inadequate road infrastructure. But there is ample evidence to show that banks, marketing, service centers, and other support activities, tend to be located where infrastructure is adequate (Binswanger et al., 1989). Whether or not to improve rural infrastructure services is only one kind of question to consider. Perhaps more important is the question of how rural infrastructure services should be provided and how we can establish a link between farm and non-farm sectors in the rural economy. This has been subsequently analyzed in the latter section.

MODERN FARM PRACTICES AND THEIR IMPORTANCE

A modern farm practice, in ordinary term, implies adopting a new farm practices by replacing the traditional ones (Pandey, 1989). In other words, it is a changing sign of agriculture from traditional to modern one through the conversion of new agricultural technology into production accomplishment. It is not a single component. There are multi-facet of factors that involved in the issue of modern farm practices in agriculture.

These include development of HYVs, improved methods of crop fertilization and discovery of new means of disease and pest control, etc. They not only help agriculture but also bring linkage between agriculture with non-farm sector and hence, ensures overall rural development in the economy. In agriculture, it helps in increasing production and productivity. However, the other importance of modern farm practices are increasing household incomes and reducing the incidence of rural poverty. It is, therefore, important that an optimum utilization of the farm potential of a territory is not possible unless sufficient innovatory changes are introduced in the modern techniques of agriculture (Tiwari and Jain, 1989). But it is to be noted that the components involved are not universal but are confined to limited areas, limited farmers and certain specific crops only. This is because these techniques are yield-increasing and helps in enhancing the farm productivity, if they are supported by both rural infrastructure and rural non-farm employment (Singh, 1976; Dube, 1987).

ANALYTICAL FRAMEWORK

The basic objective of this study is to examine the impact of rural non-farm employment, rural infrastructure and modern farm practices on agricultural productivity in India. But since rural infrastructure and modern farm practices involve number of items, we first integrate them into a single composite index separately. And for this, we applied Principal Component Analysis (PCA). PCA is a multivariate technique that transforms the original variables into a new set of variables (P_i) called principal components (Dutta et al., 2001; Mallikarajun, 2002; Bhakar and Bhargava, 2003). Expressing it mathematically,

$$P_1 = a_{11}X_1 + \ldots\ldots\ldots\ldots\ldots + a_{1n}X_n$$

$$P_2 = a_{21}X_1 + \ldots\ldots\ldots\ldots\ldots + a_{2n}X_n$$

$$P_m = P_{m1}X_1 + \ldots\ldots\ldots\ldots + P_{mn}X_n$$

The a(s) called loadings, are so chosen that the constructed principal components satisfy two conditions: (a) the principal components are uncorrelated and (b) the first principal component p_1 absorbs and accounts for the maximum possible proportion of the total variation in the set X's, the second principal component absorbs the maximum of the remaining variation in the X's (after allowing for the variation accounted for by the first principal component) and so on. For the present study, the composite indices are worked out with the help of first PCA only. However, the formula used for constructing composite index is as follows:

$$\text{CI} = \sum_{i=1}^{n} a_i \frac{X_{ij}}{S_{xj}}$$

Where, i = 1, 2, ..., n items; j = 1, 2, ... m states/regions; CI = Composite index; a_i = Loadings of items; s_{xj} = Standardized variables of items. The items used for constructing composite infrastructure index and modern farm index are reported in appendix.

So far as mathematical modeling is concerned, we simply use bi-variate regression analysis to analyze our objectives. The used model is as follows:

AGP = a + b RINF ....(1)

AGP = a + b RNFE ....(2)

AGP = a + b ADOPT ....(3)

RNFE = a + b RINF ....(4)

RNFE = a + b ADOPT ....(5)

ADOPT = a + b RINF ....(6)

Where, AGP = Agricultural productivity; RINF = Composite index of rural infrastructure; RNFE = Rural non-farm employment (in percentage); ADOPT = Composite index of modern farm practices; a and b are the parameters to be estimated.

The data under the present study are primarily secondary in nature and have been collected from the following sources: Centre for Monitoring Indian Economy, Mumbai; Handbook of Statistics on Indian Economy, Reserve Bank of India (Mumbai); and Economic Survey, Government of India (New Delhi).The study has been analyzed for two cross sectional years (i.e. 1983-84 and 1993-94) and that has been chosen as per the availability of data to analyze our present objectives.

RESULTS AND DISCUSSION

In this section, we first highlight the status of Indian states with respect to rural non-farm employment, rural infrastructure, and adoption of modern farm practices and their impact on agricultural productivity in the Indian economy. While their statuses are represented in Figures 2, 3, 4 & 5, the estimated results are reported in Tables 1 & 2. The results ensure that south zone and north zone are comparatively better position than that of east zone and west zone in all these four indicators. Among the states, while Punjab, Kerala, Haryana and Tamil Nadu are significantly better positions, Orissa, Madhya Pradesh, Bihar and Rajasthan are comparatively inferior position than that of others. The above comparison among the states literary reflects that the presence of rural non-farm employment, availability of rural infrastructure and adoption of modern farm practices can determine the level of agricultural productivity in India and across its states/zones. This is because the availability of rural infrastructure, rural non-farm employment and modern farm practices are very consistent with agricultural productivity. Where they high, there is high agricultural productivity and vice versa. This has also been subsequently confirmed by the statistical analysis.

The regression results reveal that rural infrastructure, modern farm practices and rural non-farm employment registered a significant positive impact on agricultural productivity in India and that's true for both the

Figure 2
The Status of Rural Infrastructure in the Indian States/Zones

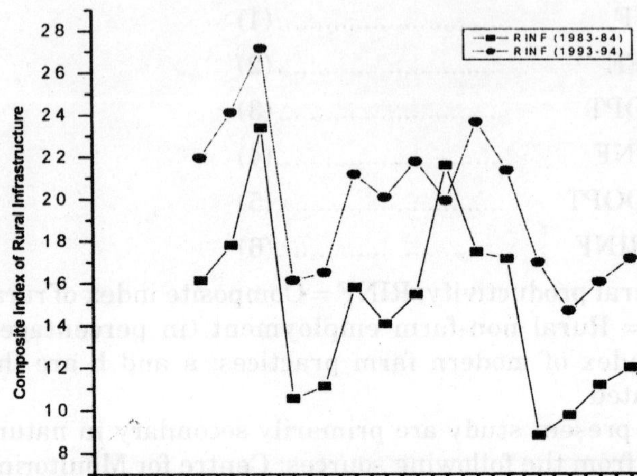

Figure 3
The Status of Rural Non-farm Employment in the Indian States/Zones

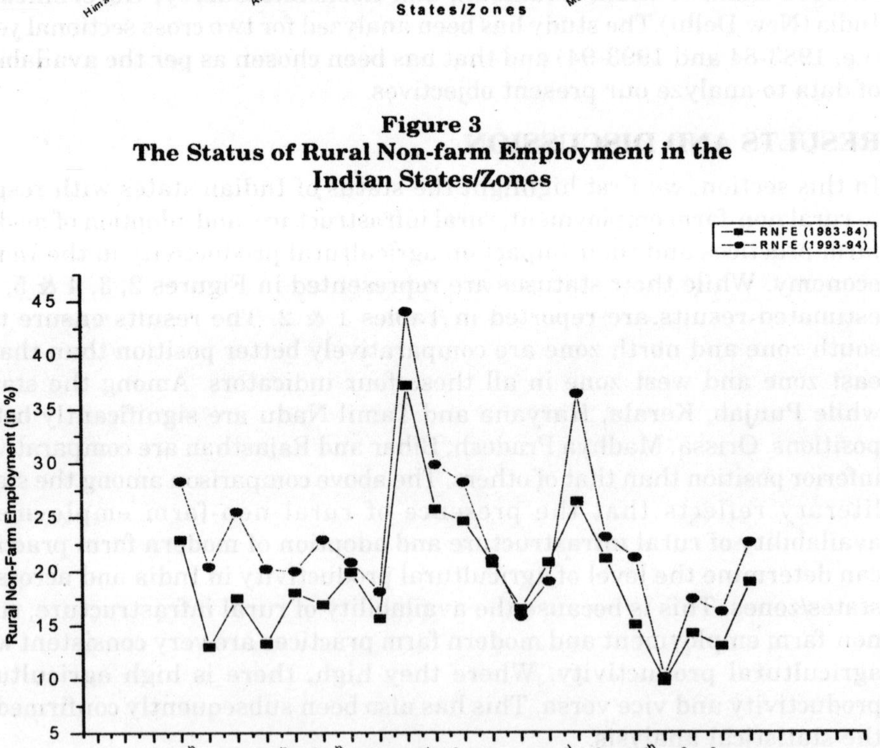

Figure 4
The Status of Modern Farm Practices in the Indian States/Zones

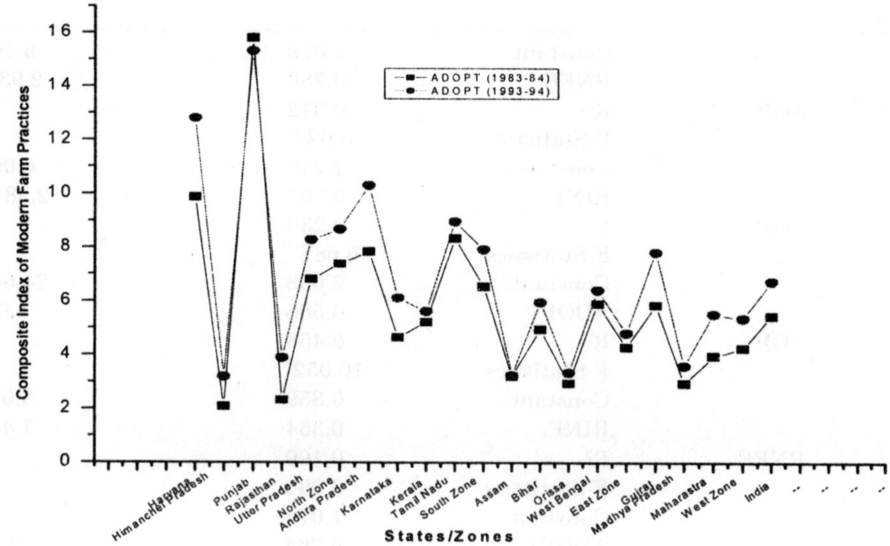

Figure 5
The Status of Agricultural Productivity in the Indian States/Zones

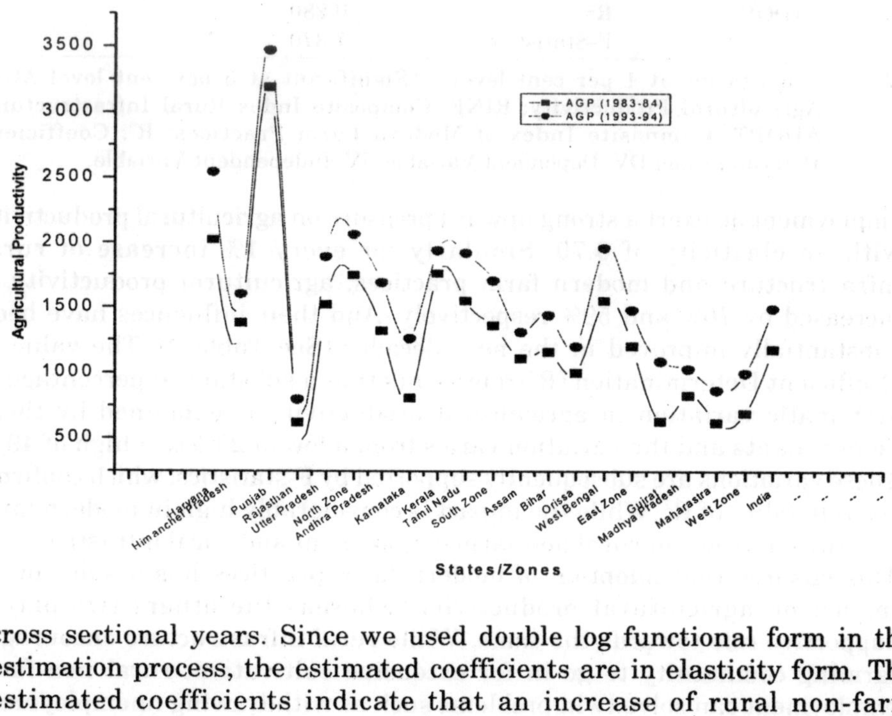

cross sectional years. Since we used double log functional form in the estimation process, the estimated coefficients are in elasticity form. The estimated coefficients indicate that an increase of rural non-farm

Table 1
Estimated Regression Results (1983-84)

Eq. No.	*IV*	*DV*	*β Coefficients*	*t-Statistics*
		Constant	2.079	6.106
		RNFE	0.789	2.937*
1.	AGP	R^2	0.312	
		F-Statistics	8.627*	
		Constant	2.210	6.061
		RINF	0.761	2.385**
2.	AGP	R^2	0.230	
		F-Statistics	5.687**	
		Constant	2.658	24.545
		ADOPT	0.588	4.01*
3.	AGP	R^2	0.458	
		F-Statistics	16.052*	
		Constant	0.859	3.075
		RINF	0.354	1.447
4.	RNFE	R^2	0.100	
		F-Statistics	2.095	
		Constant	1.074	11.42
		ADOPT	0.265	2.080
5.	RNFE	R^2	0.185	
		F-Statistics	4.325	
		Constant	-0.389	-0.959
		RINF	0.964	2.715
6.	ADOPT	R^2	0.280	
		F-Statistics	7.370	

Note: *Significant at 1 per cent level; **Significant at 5 per cent level AGP: Agricultural Productivity; RINF: Composite Index Rural Infrastructure; ADOPT: Composite Index of Modern Farm Practices; R^2: Coefficient Determination; DV: Dependent Variable; IV: Independent Variable.

employment do exert a strong upward pressure on agricultural productivity with an elasticity of 0.79. Similarly for every 1% increase of rural infrastructure and modern farm practices, agricultural productivity is increased by 76% and 59% respectively. And their influences have been substantially improved in the next decades (See Table 2). The value of Coefficient Determination (R^2) represents that a substantial percentage of systematic variation in agricultural productivity is explained by these determinants and the variation varies from a low of 23% to a high of 49%. These variations are subsequently supported by F-statistics, which confirms the reliability of R^2. While the impact is considerably high in modern farm practices, it is low in rural non-farm employment and rural infrastructure. This ensures that adoption of modern farm practices has a substantial impact on agricultural productivity, whereas the others two play a supportive role towards the same. While rural infrastructure encourages farming community to go in for extensive cultivation, rural non-farm employment can solve twin problems such as withdrawing unemployment (disguised) from the agriculture to non-farm sector and financial problems by raising the household income of the rural poor.

Table 2
Estimated Regression Results (1993-94)

Eq. No.	*IV*	*DV*	*β Coefficients*	*t-Statistics*
		Constant	0.224	7.907
		RNFE	0.701	3.33*
1.	AGP	R^2	0.368	
		F-Statistics	11.087*	
		Constant	1.667	3.589
		RINF	1.184	3.261*
2.	AGP	R^2	0.359	
		F-Statistics	10.631*	
		Constant	2.730	25.15
		ADOPT	0.565	4.248*
3.	AGP	R^2	0.487	
		F-Statistics	18.044*	
		Constant	0.218	0.506
		RINF	0.879	2.605**
4.	RNFE	R^2	0.263	
		F-Statistics	6.783	
		Constant	1.118	9.283
		ADOPT	0.279	1.893
5.	RNFE	R^2	0.159	
		F-Statistics	3.584	
		Constant	-1.045	-1.805
		RINF	1.441	3.183*
6.	ADOPT	R^2	0.348	
		F-Statistics	10.132*	

Note: All notations are defined earlier.

Whilst studying the impact of rural infrastructure on modern farm practices, the estimated results ensured that they are significantly positively related to each other. The R^2 suggests that about 28% of the systematic variation on modern farm practices is explained by the rural infrastructure during the period 1983-84 and that has been substantially improved to 35% during 1993-94. Thus, we can mention that rural infrastructure is a push factors towards the adoption of modern farm practices in agriculture. We have also studied the impact of rural infrastructure and modern farm practices on rural non-farm employment. They are all positively related to each other. The estimated coefficients indicate that for every one per cent increase of rural infrastructure and modern farm practices, rural non-farm employment increased by 35% and 27% respectively for the period 1983-84 and that has been substantially improved to 88% and 28% respectively for the period 1993-94. Though their impacts are not highly significant, but it is enough to substantiate that they certainly boost the rural non-farm employment in the Indian economy.

CONCLUSION AND SUGGESTIONS

Over and above, the study finds that rural non-farm employment, rural infrastructure and modern farm practices have substantial positive impact on agricultural productivity in India. While the impact of modern farm practices is significantly high, the other two are less impact on the same. This is probably due to the fact that they are essentially supportive factors and closely related to modern farm practices, which certainly determine the productivity in agriculture. The statistical investigation of rural infrastructure on modern farm practices also confirmed the same. The study also explored the impact of rural infrastructure and modern farm practices on rural non-farm employment. The results established that both have significant impact on rural employment. While rural infrastructure increases rural non-farm employment directly, modern farm practices can obtain the same indirectly. For instance, a substantial adoption of HYV will lead to increase the productivity of agriculture and thus, augments farm income. An increase of farm income further leads to increasing demands of consumer goods, which is more or less supplied by rural non-farm sectors. As a result, it improves the production of rural non-farm sectors and also the rural non-farm employment.

All in all, we can cite that rural infrastructure is a vital factor to bring rural development in the Indian economy. It directly affects rural non-farm employment and modern farm practices and indirectly influence agricultural productivity in particular and hence, rural development in general. As you know India is a rural economy and it can not shine without shining its rural economy. This requires a comprehensive farm and non-farm linkage in the rural economy. Without which it is very difficult to maintain rural development in the Indian economy. In this context some of the fruitful suggestions are as follows.

- There is need to improve the overall rural infrastructure base (in terms of quantity and quality) in the country and across its states/zones under a well thought out programmes of the government.
- It is required to modernize and diversify agriculture by applying modern methods of production. This is since it has very close link with the growth of agriculture and rural non-farm sector.
- There is urgent necessity to boos rural non-farm sector, which can help rural economy in various angles such as absorbing excess agricultural labour, raising household incomes, controls rural-urban migration, use of local talents, etc.
- There is need to popularize the importance of rural infrastructure, rural non-farm sector and modern farm practices particular for the disadvantaged states of India.
- Last but not the least is to encourage people's participation, which has substantial impact on all spheres of rural development.

REFERENCES

Ahluwalia, M.S. (1978), Rural Poverty and Agricultural Performance in India, *Journal of Development Studies* 14 (3): 289-324.

Arnon, I. (1981), Modernization of Agriculture in Developing Countries: Resources, Potentials and Problems, John Wiley & Sons Publications, Binghamton.

Bhakar, R.R. and Bhargava, P. (2003), "Disparities in Infrastructure Development in Rajasthan" *Indian Journal of Regional Science*, 35 (1): 57-66.

Binswanger, H., Khandker, S. and Rosenzweig, M. (1989), "How Infrastructure and Financial Institutions Affect Agricultural Output and Investment in India", World Bank Working Paper, No. 163, Washington D.C.

Dev, S.M. (2006), "Policies and Programmes for Employment", *Economic and Political Weekly*, 41 (16): 1511-1516.

Dube, R.S. (1987), Agricultural Geography: Issues and Applications, Gian Publishing House, New Delhi.

Dutta, P.C., Choudhury, B. and Roy, N. (2001), "Identification of Backward Districts for The Regional Agricultural Development- A Case Study of Assam", *Indian Journal of Regional Science*, 33 (1): 14-21.

Datt, G. and Ravallion, M. (1998), "Why Have Some Indian States Done Better than Others at Reducing Rural Poverty", *Economica*. 65 (1): 17-38.

GOI (1999-2000), Economic Survey, Ministry of Finance, Government of India, New Delhi.

GOI (2001), "National Agricultural Policy", *Yojana*, 45 (1): 4-11.

Hshim, S.R. (2004), "Indian Agriculture: The Need for a New Direction", *Agricultural Economics Research Review*. 17: 1-10.

Lanjouw, P. and G. Feder (2001), "Rural Non-farm Activities and Rural Development: From Experience towards Strategies", World Bank Working Paper, No. 22, Washington DC.

Mallikarajun, M. (2002), "Intra Regional Disparities in Economic Development- A Study of Andhra Pradesh", *Indian Journal of Regional Science*, 34 (1): 1-11.

Pandey, K.N. (1989), Adoption of Agricultural Innovations, Northern Book Centre, New Delhi.

Pradhan, R.P. (2005), "Boosting Rural Employment in a Globalized Regime", *Social Change*, 35 (4): 69-82.

Radhakrishna, R. (2002), "Agricultural Growth, Employment and Poverty: A Policy Perspective", *Economic and Political Weekly*, 37 (3): 243-250.

Ranganathan, V. (2003), "World Bank and India's Economic Development", *Economic and Political Weekly*, 38 (3): 236-241.

Ravallion, M. (1991), "Reaching the Poor Through Rural Public Works", The World Bank Research Observer, No. 6, July.

Sen, R.K. (2003), Globalization of Food Items, FDI, and Human Welfare: A Case Study for West Bengal (India). In: Kalirajan, K. and Sankar, U. (Eds.), *Economic Reforms and the Liberalization of the Indian Economy*, Edward Elgar, Massachusetts, 124-135.

Singh, J. (1976), An Agricultural Geography of Haryana, Kurukshetra, Vishal Publication, New Delhi.

Singh, P. (2002), "Realizing an Agricultural Dream", in *The Hindu Survey of Indian Agriculture*, Kasturi & Sons Ltd., Chennai, 15-22.

Tiwari, P.D. and C.K. Jain (1989), Modernization of Agriculture and Food Availability in India, Northern Book Centre, New Delhi.

Ut, T.T., Hossain, M. and Janaiah, A. (2000), "Modern Farm Technology and Infrastructure in Vietnam", *Economic and Political Weekly*, 35 (52): 4638-4641.

WDR (1994), World Development Report, Oxford University Press, New York.

APPENDIX

The items covered under the construction of rural infrastructure index are:

Irrigation: Gross Irrigated area (GIA) as a percentage of Gross Cropped Area (GCA)

Electrification: Percentage of village electrification

Transport: Road length per 100 sq. km.

Railway length per 100 sq. km.

Post office per lakh population

Health: Primary health centers per lakh population

Hospital and dispensary beds per lakh population

Education: Primary schools per lakh population

Middle/ Higher schools per lakh population

Percentage of rural literacy

Finance: Number of financial institutions per lakh population

Per capita credit to agriculture

The items covered under the construction of modern farm index are:

HYV: HYV coverage as a percentage of area under food crops

Fertilizer: Consumption of chemical fertilizers (Kg/ha)

Pesticides: Consumption of pesticides (kg/ha)

Machinery: Number of tractors (as a proxy for all machinery) per 1000 hectare of NSA

31

ENVIRONMENTAL CONCERNS IN AGRICULTURE: A FACTOR-PRODUCT AND FACTOR-FACTOR ANALYSIS

Vasudev Sedam H. & Mohammed Younus K.

SECTION - I

INTRODUCTION

Sustainable agriculture is a mode of farming that attempts to provide *long-term sustained yields*, through the use of *ecologically sound management technologies*, such as Crop diversification, organic soil management and biological pest control (Altieri-1994). The time period attached to sustained yields and the necessity of agriculture management technologies to be ecologically acceptable are important parameters in the scheme of sustainable agriculture which are considerably different from what, at present, the farmers in India, especially, in the irrigated regions of the country, are practising. The sustainable agriculture dependes much on the ability of the soil to function in its immediate environment for particular use and interacts positively with the general environment. To exploit soil's potential continuous flow of water, i.e. irrigation, addition to the fertility of the soil through the application of synthetic fertilizers and pesticides have been major input-factors and also in initiating agricultural modernization in the country. It also means maximum extraction of the quality of the soil to strengthen a standing crop and increase its yield in a given period of time. This type of human intervention has damaged the sustainable quality in the agriculture sector, reflected in the stagnancy in the production levels especially, of those crops which represented the cause of Green Revolution like, rice, wheat and cotton. The input-factors are increasing, yet, aggregate output of product has not increased proportionately (Swaminathan-2003). On the other side single type of synthetic fertilizers, pesticides and intense use of irrigation are the most important components of production function in India, i.e. there has been no innovation in production function. It also means that the ecological environment of the soil has received a serious damage due to synthetic inputs and inundating irrigation practices followed so far.

It can be sensed here that the environmental concerns in agriculture are directly related to the use of inputs. Except condition of land-size optimality, inputs in agriculture are perfectly divisible, including labour-divided into time-periods. A higher yield is achieved through optimal combination of factors of production. The optimal combination of factors is arrived through factor substitution until the amount of each factor changed (increased/decreased) does not have any negative impact on production yields. Therefore, the problem of environment threat and stagnancy is related to the models of both factor product and factor-factor. It is also necessary to bear in mind that stagnancy in agricultural production and environmental risks are intertwined. This paper seeks to analyse a specific problem of a relationship between environmental degradation and production slump in agriculture.

SECTION - II

AGRICULTURE, MILKING THE NATURE

The primitive agriculture was a production function of the human efforts with land, i.e.,

Agriculture Production *(AP)* = Land *(L)*+ Efforts(E), formally it is written as,

$$AP = f(L, E_L) \quad \text{.... (1)}$$

In this equation effort includes, primarily labour (L) devoted to tilling, sowing, protecting and harvesting. It all indicates a manual and an organic job. That is, harnessing the nature from its own resources and protecting from its own threats with means available within itself. The application of genetically modified seeds (Gs), synthetic fertilizers (F), pesticides (P), irrigation (I) and farm mechanization (M) is what constitutes agricultural modernization. Therefore, the equation (1) is rewritten to,

$$AP = f(Gs, L, E_L, F, P, I, M) \quad \text{.... (2)}$$

There is a long standing tradition in economics that states, diminishing returns very aptly apply to land and its product. The rent was determined on the basis of the quality (*Q*) of land (Ricardo-1911) If, the quality variable is included in the equation (1) it would appear like,

$$AP = f[(Q_T \cdot L) \times (E_L)] \quad \text{.... (3)}$$

The subscript '*T*', time period, in equation (3) makes the model dynamic as it is accepted that repeated use of a piece of land changes its quality. How the soil's quality is measured is a question who is measuring it. For soil analyst it could be its chemical composition of the soil and for a farmer, the yield from a given piece of land. Though we consider both, for the sake of argument in this paper we are at farmer's side. Nevertheless, the quality of land in agriculture is its fertility, which is the ability of the soil to provide fourteen essential nutrients which as nitrogen, phosphorous, copper etc. in proper balance to plants (Colvin-1994). The input factors mentioned in the equation(2) may be supposed to add to the fertility of the soil when there is a lack of it or to speed the plant growth. The production function thus transforms to,

$$AP = f\,[\,(Q_T \cdot L) \times (Gs, L, E_L, F, P, I, M)\,] \qquad \text{.... (4)}$$

The equation (4) means that the agricultural production would change with the difference in the relation to the total sum of other inputs. Since, we are at the farmers side, we would take yield from a given piece of land in a given period as the measuring parameter of quality. The implication of such an equation is that no matter what is the role of other inputs, if the change in yield is negative or zero then it needs to be concluded that the fertility of the soil is decreased or the soil is dead and that, there is a need for alternative mode of production. Since, land (i.e., soil) is indispensable to agriculture for sustained production levels new land is brought under the plough or the fertility of the same land is revitalized through an alternative set of input factors. Since we have equated quality of land to the change in yield ($\acute{y}$) a decreased AP would mean that the fertility of the soil is altered negatively either by the quality or the quantity of input factors used. Therefore, the equation (4) will be as follows,

$$AP = f\,[\,(\acute{y}_T \cdot L) \times (Gs, L, E_L, F, P, I, M)\,] \qquad \text{.... (5)}$$

Here, yield is considered in a series of time periods. It needs to be further emphasized that the each input correlates to the yield and the land variable. The sum consequence of the equation is agricultural production.

This implies that the factor-inputs can be substituted between them until a desired yield ($\acute{y}_T$) in a time period is achieved. Now, which input-factor needs to be substituted depends on its importance in determining an expected yield. That is, if factor elasticity of yield is greater, then it is vital for sustained yields. If it is vice-versa in case of an input-factor then it can be completely dispensed with or at least used sparingly. In case of greater factor elasticity of yield, an alternate factor is substituted if, environmental harms are overwhelming in the presence of certain factor/s. It is similar even with a set having input-factors which are highly complementary to each other.

SECTION - III

FACTORS IN GREEN REVOLUTION

The success of the new strategy in agricultural development-broadly called the Green Revolution depends on its package; high yielding variety (hyv) seeds, fertilizers, controlled water supply, insecticides and mechanical equipment are the main components (Chopra-1986). The degree of modernization in agriculture as a concept clearly relates to the use of package of improved complementary factor inputs associated with the modern farm technology. As a result of modernization in agriculture output of agricultural produce has increased manifold and this is a common knowledge. The success story of green revolution is rife with record breaking production of agricultural crops (Buck *et al.*,-1967, Singh-1967, Vaidyanathan-1987). There has been increase in employment (Bhalla-1987) in agricultural sector. The income in agriculture sector has increased but the distribution has been unequal (Junankar-1975, Narayan-1990).

However, the studies which are with respect to input factors in agriculture are but for the purpose to show that these factors are positively related to increase in output. But, the studies on the externalities in the form of damage done to soil and other environment implications are less and far in between in economic literature. For example, rice, wheat, and cotton crops have been the most genetically modified and extensively tried in green revolution. It is a common sight in irrigated areas that rice, wheat, sugarcane, oil-seeds or cotton are grown more than any other crop. Complementary input-factors to irrigation in agriculture modernization are chemical fertilizers, and pesticides. Therefore, we can assume that these crops not only represent gains of modernization but also the possible externalities of chemical input factors which can also be taken account of (Rao-1975). These are the crops which have undergone the stress of modernization more. And hence, the trends in their production makes sense of general agricultural scenario, whether it is concerned with supply of food-grains and commercial crops or their productivity levels in correlation to input factors like land, hyv-seeds, irrigation, and fertilizers. For example, the table-1 depicts a peculiar scenario. This data and in the following tables pertains to the agriculture sector in Karnataka, India. At the general level irrespective of the crops the consumption of fertilizers is increasing. The reason is clearly because 1) more crops are coming under chemical fertilizers treatment and 2) the areas where irrigation is available the consumption of fertilizers has increased not because more land is covered under irrigation but because the irrigated lands now require larger quantity of fertilizers to let the fertility extracted from them (Colvin-1994). A simultaneous

Table 1
Factor and Produce

Year	*Fertilizer (LT)**	*Irrigation (LH)#*	*Hyv-seeds (LH)*	*Rice Kgs. / H@*	*Wheat Kgs / H*	*Cotton Kgs / H*
1993-94	8.05	24.07	31.64	2439	852	242
1994-95	8.23	24.26	40.36	2574	761	229
1995-96	9.36	25.02	36.03	2516	701	254
1996-97	8.26	25.43	39.46	2488	807	274
1997-98	11.09	25.93	49.44	2499	198	226
1998-99	11.49	26.31	50.80	2698	829	275
1999-00	12.71	26.84	54.25	2699	877	218
2000-01	12.94	27.63	51.05	2730	988	277
2001-02	12.49	28.63	50.51	2401	803	180
2002-03	10.99	29.35	50.53	2179	630	151
2003-04	9.19	20.09	50.91	2500	436	149
2004-05	12.19	30.92	51.57	2853	779	236

Source: Karnataka Economic Survey-2005-06.

*Lakh hectares; @Kilograms per hectare.; # Lakhs in hectares.

increase in fertilizers, land under irrigation and use of hyv-seeds points to the fact that these input-factors are vital for the development of agriculture and more so, indispensable to increase the production levels in the agricultural sector (Khan-1992). However, if we look at the yields of the representative crops, we will find that the yield per hectare has been fluctuating more often. Also take a look at the mean and deviations of the input-factors and yields of the representative crops (Table 2). In case of input-factors the deviations from mean have been high, that is, the quantity of input-factors used is being increased every successive year. Whereas, the yield have remained consistently stable or static with high fluctuations (Table 1).

Table 2
Deviations of Input-factors & Yield*

Particulars	*Mean*	*Standard deviation*	*Standard errors*
Fertilizer	10.5825	1.86667	0.53886
Irrigation	26.20667	2.83265	0.81771
HYV	46.37917	7.40331	2.13715
Rice	2548	178.65813	51.57416
Wheat	721.75	214.67994	61.97276
Cotton	225.91667	44.91296	12.96526

*Based on Table 1.

The reason for such an inconsistent performance of the main crops could be only because of failure of input-factors to extract maximum yield from the soil. In other words, soil has been exploited to the maximum possible, and that, the fertility has diminished. Therefore, the factor-produce correlation has reached the static state. And, that, there is a greater need for an alternative system of crop cultivation with the same yields. As we have said in the equation (5) it is the yield, i.e., the quality of the soil, which is sole responsible variable for the agricultural production unless, the inputs used are consistent with the changing quality of the soil, which is understood by the yearly trends in the crops yield. In the beginning itself we had pointed-out that, notwithstanding the quality of input-factors or their quantity, agriculture sector is subjected to the law of diminishing returns. For example, for the total production of rice, wheat and cotton in Karnataka, the area of land used is more or less same (Table 3).

With negligible deviations from mean (figures in parentheses of the table 3) area of land under these crops has not increased total crop production. In fact, on an average the production of rice, wheat and cotton have seen deep slumps with negative growths compared with their respective average production figures. It could be argued that if the drought and other climate distortions are accounted for the negative growth the production levels might be reasonable. But, in Karnataka, we scarcely have had any flood. The only natural calamity in the state is drought. However, the selection of the crops is such that these are grown mainly in the irrigated

areas which are to an extent immune to the implications of drought. It also makes sense that, in drought prone areas these crops are grown less and the use of hyv-seeds, fertilizers and pesticides is associated mostly with agricultural modernization in the areas irrigated. It is, therefore, clear that the environmental concern in the agriculture sector is established more strongly on the basis of input-factors used and the production levels.

Table 3
Area and Production of Rice, wheat and Cotton
(Area=Lakh Hectares : Production = Lakh Tonnes)

Year	*Rice*		*Wheat*		*Cotton*	
	Area	*Production*	*Area*	*Production*	*Area*	*Production*
1993-94	13.74 (2.03)*	31.83 (-0.91)	2.37 (-0.27)	1.92 (0.06)	5.71 (0.21)	7.10 (-0.19)
1994-95	12.96 (1.25)	31.67 (-1.07)	2.44 (-0.2)	1.76 (-0.1)	6.36 (0.86)	8.15 (0.86)
1995-96	12.65 (0.94)	30.24 (-2.5)	2.19 (-0.45)	1.46 (-0.4)	6.78 (1.28)	9.63 (2.34)
1996-97	13.59 (1.88)	32.12 (-0.62)	3.48 (1.16)	1.90 (0.04)	6.60 (1.10)	10.09 (2.8)
1997-98	13.53 (1.82)	32.13 (-0.61)	2.51 (-0.13)	1.18 (-0.68)	5.02 (-0.48)	6.34 (-0.95)
1998-99	14.27 (2.56)	36.57 (3.83)	2.69 (0.05)	2.19 (0.33)	6.36 (0.86)	9.77 (2.48)
1999-00	14.50 (2.79)	37.17 (4.43)	2.61 (-0.13)	2.18 (0.32)	5.46 (-0.4)	6.65 (-0.64)
2000-01	14.83 (3.12)	38.46 (5.72)	2.66 (0.02)	2.50 (0.64)	5.52 (0.02)	8.55 (1.26)
2001-02	14.18 (2.47)	32.34 (-0.4)	2.60 (-0.04)	1.99 (0.13)	6.08 (0.58)	6.12 (-1.17)
2002-03	11.55 (-0.16)	22.37 (-10.37)	2.47 (-0.17)	1.60 (-0.26)	3.93 (-1.57)	3.67 (-3.6)
2003-04	10.74 (-0.97)	22.50 (-10.24)	2.32 (-0.32)	1.88 (0.02)	3.17 (-2.33)	4.96 (-2.33)
2004-05	13.08 (1.37)	35.47 (2.73)	2.42 (-0.22)	1.79 (-0.07)	5.12 (-0.38)	6.88 (-0.41)
Total	140.62	392.83	31.76	22.35	66.11	87.51
Mean	11.72	32.74	2.64	1.86	5.50	7.29

Source: Karnataka Economic Review-2005-06.
*Figures in parentheses are mean deviations.

SECTION - IV

ENVIRONMENTAL CONCERNS: FACTS AND FACTORS

Having made ourselves acquainted with the aspects in agricultural production function revealing the pressure on soil fertility exerted by the modern chemical inputs and modern cultivation method, we can move forward to discuss some of the important facts and factors which show how critical it is to find an alternative set of input-factors which are

environmental friendly. We have, indicated above how input-factors can be judged with the help of changes (increase or decrease) in the yield of agricultural produce. However, it is a fact that the intense external-input oriented agriculture, which was part of our green revolution strategy, has depleted soil fertility considerably in all our major agricultural production systems. This is reflected in the declining trend of total production from the same land, diminishing response of food grains, increase in the application of fertilizer nutrients, reduction in the organic matter content of fertile soils, and increase in the productivity constraints to land such as ground-water depletion, salinity, acidity alkalinity and the like (Raman-2005). The results from long-term fertility experiment at various places in the states like West-Bengal, Andhra Pradesh and Karnataka indicate a decline in organic matter content of soil with the application of recommended doses of inorganic fertilizers (Nambiar-1994). A study points out that, among fertilizers, the conversion of fertilizers to gaseous forms–ammonia and various oxides of nitrogen leads to atmospheric pollution. Escape of fertilizer as ammonia gas is called ammonia volatilization. The presence of ammonia and sulphur-di-oxide may lead to acid rains which ultimately degrade the soil (Societies-2004). The continuous use of irrigation has resulted in water logging and consequent salinity of soil (Sarkar-2001). Besides, the heavy use of pesticides and insecticides, as a part of modern crop-management system, is reported that it might alter certain characteristics of crop plants, usually, taste and nutritive contents (Mazamorasch-1988).

The facts stated above are directly related to the sustainability of yields. At the most basic level, environmental sustainability can be presented as a function of five phenomena (Esty-2002):

(1) the state of the environmental systems, such as air, soil, ecosystems, and water;

(2) the stresses on those systems, in the form of pollution and exploitation levels;

(3) the human vulnerability to environmental change in the form of loss of food resources or exposure to environmental diseases,

(4) the social and institutional capacity to cope with environmental challenges and finally;

5. the ability to respond to the demands of global stewardship by co-operating in collective efforts to conserve international environmental resources such as the atmosphere.

However, these are broad parameters. The human vulnerability to shortages of food supplies has had lead to malnutrition, especially, among women and children and, in the past, famines.

In the Environmental Sustainability Index under its components, in environment systems land is said to be sustainable in terms of quality[1] and fertility[2] if there is:

(a) reduction of water stress and

(b) reduction in fertilizer consumption and pesticide use per hectare of crop land.

Recently, in the present crisis, an alternate fertilizer nutrient in the form of integrated use of organic yield booster is a major concern for efficient modern type of land cultivation. There are reports which suggest that the combined use of inorganic fertilizers and organic manures enhanced the inherent nutrient-supplying capacity of the soil and improved the physical and biological properties of soil, which, in turn improved the grain yield (Hussain-1995). Green manuring shows an increasing trend of soil quality; whereas with inorganic fertilizers (NPK) the soil organic carbon content decreases drastically even with laboratory standard NPK soil treatment (Hali-2001). There are also studies which say that for sustainability of an agricultural system, it is desirable not to grow a particular crop of a group of crops on the same soil layers for meeting their water and nutrient requirements, depending on their root and shoot systems (Sing-2001).

All the above facts point towards a possible set of alternative input-factors that would help create sustainable agriculture system. The following schemata have been suggested for factor-factor substitution (Altieri-1994). The input-factors used under the modern agricultural system has been labeled as *conventional* and its substitute for sustainable system as *alternative*.

Factors	*Alternative*	*Conventional*
Sun	Man source of energy	Energy used as "catalyst" for Conversion of fossil energy
Water'	Mainly rain & small irrigation schemes	Increased use of large dams and centralized water distribution system
Nitrogen	Collected from air by legumes & recycled	Primarily from synthetic fertilizer
Minerals	Released from soil reserves & recycled	Mined, processed & imported
Weed & Pest control	Biological & Mechanical	With inorganic pesticides
Energy	Some generated & collected on farm	Dependence on fossil fuel like petrol & diesel
Seed	Some produced on farm & Local	All purchased hybrids
Management decision	By farmer and community	Provided by agri-business
Animals	Produced supergistically on farm	Feed lot production at Separate location
Cropping systems	Rotations & diversity enhance value of all of above components	Monocropping
Labour	Most work done by the Family living on the farm	most work done by hired labour
Capital	Initial source is family and community, any accumulation wealth is re-invested locally	Initial source is external indebtedness or equity & any accumulation flows mainly to outside investments

The above scheme for alternative farm cultivation is based on input-factors from within the nature, though in some cases it may also appear that the scheme is capital and labour saving. However, as far as maintaining the sustainability of land fertility and quality is concerned the question of capital and labour is external, i.e. may not be considered as the organic parts of the land cultivation in terms of sustainable agriculture practices. However, the large scale production might definitely require the use of farm machinery and extra labour. But then, it is determined by the land size under the possession a farmer and the time constraint. As it is already mentioned in equation (5) that the importance of an input factor completely depends on how it is correlated with the yield (or quality). If the trend in the yield is positive with each successive set of input-factors in each time period then the use of that set of inputs is retained in successive applications in land cultivation. Till then it takes a rigorous factor-factor substitution in relation with the desire for high production levels. A study sponsored by National Academy of Agriculture Research Management, Hyderabad, India has forecasted the sources of sustainable productivity in agriculture. In the study the technology used is directly related to the type of input-factor is used and its deterministic role in crop yield (Raman-1988). The technologies and their importance in percentage terms has been shown in the following table:

Table 4

Technologies and their due in Land productivity

Technologies'	*Due in the Productivity increase (In percentage terms)*
Plant breeding	135
Irrigation and crops to conserve water	133
Genetic regulators	125
Growth regulators	124
Rising level of atmospheric carbon dioxide	120
Biological nitrogen fixation	118
Multiple cropping and polyculture	115
Improved photosynthetic efficiency	117
Temperature acclimation	113
Forage nutritional quality	112
Crop maturity	111
Transpiration suppressants	110
Intercropping	108
Protected cultivation	105

Source: Raman-1998. "Agriculture Research system Management in the 21st Century", National Academy of Agriculture Research Management, Hyderabad.

A close comparison of sustainable agriculture scheme and the inputs importance forecast shows that the organic input-factors have received greater importance right from the beginning. For example plant breeding, genetic regulators, biological nitrogen fixation, multiple cropping & polyculture, forage nutrition, intercropping in table 4 are contingent upon the agro-ecological practices in the Alternative schema, like, using energy from sun, collecting legumes to supply nitrogen to crop plants, using soil resources to supply minerals, using farm produced seeds, cropping diversity and intercropping. A deep analysis of the facts and factors described above insinuates us to call back the traditional form of agriculture system wherein the factors determining agricultural practices came from nature itself. Although, a bit of human intervention might also be conceded after noting the fact that genetically modified seeds have played a major role in increasing the production levels. And as we have said these hyv-seeds when used with ecologically acceptable input-factors should respond with the same yield.

SECTION - V

CONCLUSION

The factor-product and factor-factor application in agricultural sector reveals the deterministic role of input-factors in relation to the production levels and yield trends. Land being the organic input-factor, for desirable levels of output, with expected yields per hectare input-factors, complementary in agriculture production function, needs to be ecologically acceptable. We have learned that the chemical input-factors so far used have not helped in maintaining successive levels of growth rates in agriculture sector. Factor-factor analysis has shown the necessity to look for alternatives within the complementary set of input-factors. However, we have not discussed here the production and cost implications of ecologically acceptable factors substituted for chemical or inorganic input-factors. Nevertheless, in section IV it is stated that the social and institutional structure defines the capacity of a community or nation in taking the challenges of environment sustainability. This applied to agriculture sector, it transforms to the need for emphasis on research and development related to agriculture production and yield sustainability. Most importantly, the dissemination of information about the sustainable tillage practices stands out to be most crucial.

The stagnancy in agriculture sector, is also a pointer not only towards environmental concerns based on input-factors used, but also the need to practice sustainable crop cultivation to maintain the production level proportionate to the increasing population in the country, especially, in a situation where the cases of deaths due to malnutrition are increasing. Finally, in recent times there have been reports saying that because of the loss in soil fertility farmers who failed to realize the desired level of production volumes committed suicide as they became incapable of paying their debts to their lenders. This much is enough to understand the gravity of the situation that arose due to unhealthy production practices in agriculture.

REFRENCES

Altieri, Miguel A. (1994), "Sustainable Agriculture". Encyclopedia of Agriculture Science, edited by Charles J Arntzen, Academic Press, San Diego, California, p. 101.

Burk, T.L., C. Clark, and M. Haswell (1967), "The Economics of Sustainable Agriculture" McMillan, London.

Chopra,R.N. (1986), "Green Revolution in India" IBM & Oxford University Press, New Delhi. p.——

Colvin, I.S. (1994), "Soil Management", Encyclopedia of Agriculture Science, edited by Charles J. Arntzen, Academic Press, San Diego, Caligornia, p. 109.

Esty, Daniel, Peter K. Cornelius (2002), "Environmental Performance Measurement: The Global Report 2001-2002", World Economic Forum, Oxford University Press, New York, p.14.

Hali, K.M., K.G. Mandal, A.K. Misra, P.K. Ghosh and C.L. Acharya(2001), "Effect of Irrigation Regimes and Nutrient Management on Soil Water Dynamics, evapo-transpiration and yield of wheat (Triticum Aesticum) in vertisol", Indian Journal of Agricultural Sciences, Vol. 71, No. 9 p. 583.

Hussain, Majid (1976), "A New Approach to agricultural Productivity in the Region of the Sutlej, Ganga plains of India", Geographical Review of India, Vol. 38, No. 3.

Khan, Shafiqa (1989), "Agricultural modernization in India", Anmol Publications, New-Delhi.

Mazamorasch, Karl (1988), "Is Biotechnology an answer to our future needs ?", in Agricultural Research Systems and Management in the 21st Century, edited by K.V. Raman *et al*, National Academy of Agricultural Research Management, Hyderabad.

Nambiar, K.K.M (1994), "Soil Fertility and crop Productivity under long-term Fertilizer use in India", Indian Council of Agricultural Research, New Delhi.

Narayan, D. (1990), "Agricultural Growth and Changing Structure of Assets of Rural Households", in Agriculture and Rural Economic in India", Deep and Deep Publicatios, New Delhi.

Raman, K.V., M.M. Anvar, K.B. Gaddagimath (1988), "Agriculture Research System Management in the 21st Century", National Academy of Agriculture Research Management, Hyderabad, pp. 59.

Raman, K.V. (2005), "Sustaining Soil Fertility", in Survey of Indian Agriculture 2005 The Hindu, p. 165.

Rao, Hanumanth (1975), "Technological Change and Distribution of Gains in Indian Agriculture, Delhi.

Ricardi, David (1911), "Principles of Political Economy and Taxation" Everyman's Library, London.

Singh, Jasbir (1967), "The Measurement of Agricultural Productivity", Economic Geography, Vol. 43.

Societies, Internal Federation of Red Cross and Red Crescent (2004), "World Disaster Report", Geneva, p. 63.

Vaidyanathan, A. (1987), "Irrigation and Agricultural Growth", Indian Journal of Agriculture Economics, Vol. 42.

32

TOWARDS ORGANIC FARMING

I. Satya Sundaram

In India, agriculture and allied sectors are characterised by low and volatile growth rates. In the Eight Plan (1992-97), this sector achieved an annual growth rate of 4.7 per cent, but this declined to 2.1 per cent in the Ninth Plan (1997-2002). The growth raes were (all provisional figures) 10 per cent in 2003-04, 0.7 per cent in 2004-05 and 2.3 per cent in 2005-06.[1]

Productivity in the agricultural sector continues to be low. For instance, wheat productivity in 2002 was 2770 kg/ha, but this was high at 8043 kg in U.K., 7749 kg in France and 3885 kg in Chine.[2]

The green Revolution in India has no doubt contributed to self-sufficiency in respect of food-grain output. However, in recent years, there has been no improvement in the levels of foodgrain output and yields. Agriculture scientists say that the country requires the second phase of green revolution in ensure steady growth of foodgrain output and food security.

Agriculture yields have reached a point of fatigue despite increased fertiliser use.[3] There are serious limitations to the seed-fertiliser technology. It has not been able to solve problems like low yields and crop imbalances. The ecological dimensions of the new agricultures technology need to be carefully evaluated. Hence, the current stress on organic farming.[4]

As the growth rates of yields under seed-fertiliser technology reach a plateau, the tools of bio-technology offer possibilities for breaking these barriers, especially in the unfavourable agricultural regions.[5]

This paper has been divided into two sections : (i) Advantages of Organic Farming (ii) Organic Farming: Measures Needed.

I

ADVANTAGES OF ORGANIC FARMING

Organic farming depends on a compost in the form of mixture of leaves, manure and plant waste. It is a system in which the maintenance of soil fertility and the control of posts and diseases are achieved through the enhancement of biological processes and ecological interaction.

In organic farming, oil cakes and oil meals play a key role as natural fertilisers. Repressed and mustard neem, castor, mahua, karaja and linseed cakes are commonly used as organic nitrogenous fertilisers.

Organic farming differs from commercial farming mainly in crop rotation, fertiliser applications and pest control methods.

Dependence on inorganic manure can be further reduced if green maturing can be practised by introducing a suitable short duration crop for the purpose.

Under organic farming, instead of fertilisers, a compost in the form of a mixture of leaves, manure and plant waste is used. For repelling pests, pesticides made from herbs and trees are used. Neem-based pesticides have become popular.

Crop rotation is important. Different crops obtain nutrients from different depths of soil and as such soil nutrients are utilised in a balance way. Also, recycling of crop residue máy be done by burrying the farms waste (leaves, stem etc left after the harvesting) in the field itself.[6]

The organic residue from biogas plants is an excellent source of organic matter for soils.[7]

Farm forestry has to be encouraged. Incorporating of trees in the field lands and village common lands etc. has many advantages. Farm forestry can also reduce soil erosion.

Vermi-composting is the process of growing earth-worms in various kinds of waste material and using the excretion of the earthworms as manure which contains NPK. The earthworms are useful in land reclamation, soil improvement and organic waste management.

The earthworms are environment-friendly. They take up into their tissues a number of unwanted organic and inorganic chemicals. They enrich the soil with oxygen and encourage aerobic micro organisations. They enhance water-holding capacity of the soil.[8] organic farming is cost effective.

The advantages of organic farming are many, but the following deserve mention:

- organic fertilisers provide readily available plant nutrient in a concentrated form and generally they do not have any significant effect on the physical properties;
- organic manures are low in nutrients but play a vital role in the improvement of physical, physico-chemical and biological soil properties;
- use of organic manures alone will result in a soil with improved physical and micro biological properties, use of mineral fertilisers alone will increase the quantity of available plant nutrients and the use of both together will give a soil rich nutrients with physical and microbiological properties.

Super fine rice variety 'J-18' developed at the 'J-Farm', 35 km from Chennai, has done exceedingly well in farmers' holding in Kancheepuram district under rich organic nutrition. The plant protection expenses are also minimal. The yield is as high as 4427 kg per hectare. The straw reaped is also huge in quantity. This variety grew only on organic manure and green manure. The fine variety will fetch a premium price in the market.[9]

II

ORGANIC FARMING: MEASURES NEEDED

Realising the importance of organic farming, the Government of India launched a National Programme for Organic Production and Farming in 2000.

The success of organic farming depends on the efficient agronomic management to stimulate the productivity of the soil resource.

We should give importance to Integrated Nutrient Management (INM). INM is a flexible approach to minimize the use of chemicals, but maximize use efficiency and farmers' profit. Today we find overuse and abuse of fertiliseres.

INM is based on three principles:

- Assessment of the basic soil fertility and climate;
- Nature of the crop, not in isolation, but as a part of the cropping system and yield.
- At least 30 per cent of the total nutrient levels of NPK to be in organic form.

Land is heterogeneous. The stress should be on 'precision agriculture' where site-specific management practices are adopted giving due considerations to the spatial variability of land in order to maximize crop production and minimize the environment damage. There should be judicious crop management at eh micro level wherein only required amounts of inputs are applied.

An amount of Rs. 100 crore had been allocated in the Tenth Plan (2002-07) To promote organic agriculture.

The Task Force set up by the agriculture ministry to study the various aspects of organic farming submitted its Report on November 29, 2001. It recommended the establishment of a permanent national Board to promote environment friendly chemical-free agriculture.

The Task Force suggested that all ventures engaged in the production of compost, press mud, vermi-compost and other forms of substances meant to replace fertilisers and pesticides should be exempted from taxes.

Organic farming is recommended because it alone can reduce the pressure on natural resources and can minimize the pollution from the excessive use of chemicals.[10]

Effective production and market incentives must be provided to small and marginal farmers in particular to grow crops using sustainable organic methods for domestic and export markets.

Of course, the use of chemical feritlisers cannot be completely eliminated. But, these need to be used judiciously.

NOTES AND REFERENCES

1. Government of India, Economic Survey 2005-06. p. 155.
2. Ibid., p. 156.
3. Sankaram Ayala: 'Sustainable Pathways for Agricultural Productivity', The Hindu, July 26, 2001. p. BS-3.
4. I. Satya Sundaram: 'Organic Farming: Towards Sustainable Agriculture', Civil Services Chronicle', February 2001. p. 18.
5. C.H. Hanumantha Rao and Ashok Gulati : 'Indian Agriculture: Emerging Perspectives and Policy Issues', Economic and Political Weekly, December 31, 1994. p. A-158.
6. M. Anadan: 'Organic Farming for Eco-friendly Agriculture', Employment News, September 19-25, 1998.
7. Y. L. Nene: 'Sustainable Agriculture: Only Way out', The Indian Express (Science Express), May 4, 1999.
8. Radhakrishna Rao: 'earthworm: Nature's Bio-reactors', The Indian Express (Science Express), April 28, 1998.
9. The Hindu, April 5, 2001. p. BS-8.
10. C.H. Hanumantha Rao: 'Sustainable Development of Agriculture', The Economic Times, November 16, 1998.

33

PARTICIPATORY IRRIGATION WATER MANAGEMENT IN JAMMU AND KASHMIR

Falendra K. Sudan

BACKGROUND

Over the past few decades, water scarcity has emerged as a global problem. In course of time, the level of human interventions has grown to such an extent quantitatively that the need for a more informed approach to this vital natural resource endowment is being recognized and articulated all over the world (Falkenmark *et al.*, 1980). As human increasingly started to intervene in the flow of rivers (Deursen, 2000), the gradual success of these interventions resulted in the upscaling of their operational dimensions that finally led to the emergence of large dams as instruments for river basin development (Bandyopadhyay *et al*, 2002). The large-scale supply-side augmentation has been prescribed by engineers as the only way of meeting the growing water requirements of developing countries. As a result, policy–makers in India have been adhering to the idea of solving the water problem by harnessing the huge monsoon run-off in the main rivers, primarily with the help of such large interventions (Bandyopadhyay and Perveen, 2004). The proposal for the interlinking of rivers offering a combination of storage dams and transfer canals has been offered as a long-term supply-side solution to water shortages (IWRS, 1996).

The idea of interlinking rivers is described as the perfect win-win solution that will address the problems of water scarcity and floods. In contrast with the official prescription for interlinking of rivers, there is a strong opinion that India's water crisis is caused by the mismanagement of water resources, and the solution do not lie merely in supply-side augmentations. There has been no serious attempt to work out a series of area specific answers by way of local conservation and augmentation to the maximum extent possible. Recently, the drought conditions were the result of bad water management in the past and the answer lay in better resource management in the future (Iyer, 2000). Thus, the proposed interlinking of river needs to be assessed in the light of the emerging paradigm changes in the subject of water management. What is needed is a fundamental shift away from present reductionist engineering paradigm to a holistic and trans-disciplinary one. The new ways of managing water has increasingly been identified as integrated water resource management (IWRM). A higher priority should be assigned to projects that meet basic and unmet needs of water (Gleick, 1996). Following these paradigmatic shifts in notions worldwide, various other means to conserve water in stream is emphasized. In the Indian context, the old paradigm is still strongly entrenched in the official water administration. The exclusively supply-side solutions offered by the present paradigm tend to either impinge on demands by others, or cause serious damage to nature. The demand-side management is therefore slowly becoming a new paradigm for water governance.

Nigam *et. al.* (1997) makes it clear that if the precipitation available within the respective watersheds or sub-basins is harvested and conserved properly, the satisfaction of domestic water needs would not be a problem in most parts of the country. For promoting domestic water security, local level water conservation is a suitable option compared with developing large storage and long distance diversion facilities, as these often carry high financial, social and ecological costs (World Commission on Dams, 2000). The construction of dams on the Himalayan rivers cannot be undertaken by ignoring vital questions on the uncertainty associated with taking a mechanical and traditional view of development of the Himalayan rivers (Ives and Messerli, 1989). Due to the verticality and the consequent fragility of the Himalaya, large dam on Himalayan rivers will be subject to high levels of seismic hazards, since the potential for earth quakes at the plate boundary all along the Himalayan foothills is well known and widely accepted (Khattri, 1987). Keeping this in view, the newer interventions may be counter-productive.

Besides, there has been the inherent failure of policy-makers to perceive the dimensions of the upstream-downstream linkages. Such gaps have led to a major source of conflict over water use in river basins all over the world. The inequitable distribution of fresh water leads to violent inter- and intra-basin conflicts. The long-drawn Cauvery dispute is a case in point. The situation with respect to Satluz-Yamuna Link (SYL) canal is no different. Inevitably, interstate trans-boundary water-related conflicts may become inter-country. The proposal risks major confrontation with Bangladesh, which receives much of its water from the Ganga and the Brahmaputra, after they leave India. The government of Bangladesh protested to India but has so far not received any response (Vidal, 2003). Unfortunately, many plans for river basin development in developing countries are inflexible and the requisite spirit of cooperation is often lacking (Serageldin, 1995). The results are that damages to the ecosystem services or losses in downstream population have been ignored. In the case of India, the situation is more complex, with the downstream areas belonging to another country, as is evident from the recent controversies regarding construction of Baglihar dam in Jammu and Kashmir and the steps to abrogate 1960 Indus Water Treaty. Keeping the above in view, the only solution to the problem of domestic water scarcity in dry upland lies in demand management through local-level harvesting and conservation of rainfall in decentralized and sustainable manner.

India has embarked upon Participatory Irrigation Management (PIM), under which the management of some of the systems is being turned over to the Water Users' associations (WUAs). The conservation of land and groundwater resources through rainwater harvesting is another area where market failure as well as government failure has been glaring. As a solution to this, Participatory Watershed Development (PWD) is being evolved and practiced in many countries, including India, for the protection and sustainable use of land and water resources. The users' associations reflect definite institutional arrangements in which rewards and sanctions were clearly defined in order for the systems to work. Therefore, a solution that has been tried in various places is to transfer the responsibility for the operation and maintenance (O and M) of the canal systems below certain point to the local farmers. At present, most state governments in India encourage the formation of WUAs and eventually want to transfer irrigation management to irrigators (Parthasarthy, 2000). Integrated Watershed Development Project (IWDP), Hills-II, Jammu and Kashmir decision on PIM is a step in this direction. WUAs are expected to serve as channels for feedback and feed-forward information to improve the performance and services provided by O and M irrigation personnel. The initiatives of IWDP, Hills-II should be seen as the first steps towards attaining sustainable water resource management. With this backdrop, an attempt has been made to analyze the process of evolving participatory irrigation water management under IWDP, Hills-II; and to examine the impact of watershed interventions on reduction in rainwater loss and sediment yield; status of water resources and irrigation; and change in crop intensity and crop productivity.

STUDY AREA AND METHODOLOGY

Integrated Watershed Development Project (IWDP), Hills-II, Jammu and Kashmir covers two sub-watersheds in *Shivaliks,* viz. Ramnagar and Akhnoor and two sub-watersheds in *Karewas,* viz. Rajwar and Rambiyara. The present study was confined to two sub-watersheds of Akhnoor (Jammu district) and Ramnagar (Udhampur district). While Ramnagar is in the inner *Shivaliks,* Akhnoor is on its outer part. Ramnagar sub-watershed is the catchment area of Ramnagarwali *Khad* (ephermal) in the middle catchment of river Tawi. It has an area of 32,630 Ha. and is subdivided into 39 micro-watersheds. Akhnoor sub-watershed with an area of 42350 Ha. is subdivided into 37 micro-watersheds. It covers about a dozen rivulets (*nallahs*) and a large number of small nallahs originating from *Kalidhar* ridge and draining into Chenab river on the western portion of *Kalidhar* ridge, drains join Manawar Tawi which in turn also meets the Chenab river. In terms of physiographic features, there may not be 100 per cent similarity between the selected sub-watersheds but, certainly, in term of accessibility, natural resource endowment, infrastructure development and general index of socio-economic awareness, two sub-watersheds are comparable.

The primary data has been collected using well-structured and pre-tested questionnaires, participatory rural appraisal (PRA) techniques and group meetings. Whenever possible and required, secondary data sources have also been utilized. Three WUAs each have randomly been selected from the sub-watersheds of Akhnoor and Ramnagar. Therefore, a total of six WUAs have been selected with the consultation of project functionaries at sub-watershed level. The sample size has been restricted to the members of the WUAs selected under IWDP (Hills-II) in the selected sub-watersheds. A total of 15, 12 and 19 WUAs' members have been selected from village Sunetar, Johnu and Dehari respectively in Ramnagar sub-watershed and 21, 12 and 11 WUAs' members have been selected from village Chohara, Maira and Ambarain respectively in Akhnoor sub-watershed. For the collection of primary data and information on environmental and economic impact, three micro-watersheds/villages each from the two sub-watersheds of Akhnoor and Ramnagar, where a maximum number of people are known to have benefited from each project intervention have been selected. In the non-beneficiary category, three sample villages each have been selected on random basis, each 20-25 km away from the sampled micro-watersheds/ villages (*i.e.*, project area). Therefore, a total of six WUAs (three each from selected sub-watersheds) and twelve villages (six each from 'project' and 'non-project' area) have been selected for collection of primary data and information. The stratified sampling technique has been used to select villages (where WUAs have been created, as well as 'project' and 'non-project' villages). During the stratification, care has been taken to include both "forested watersheds' villages and agricultural watersheds' villages" in order to makes a comparative study. For the collection of primary data related to environmental and economic variables, a purposive sample of approximately 20 per cent of household level respondents in 'project' and 'non-project' area. The team strategy has been used to collect data and information. The content analysis technique has been used to analyze the data and information qualitatively and quantitatively (using descriptive statistics).

PIM UNDER IWDP (HILLS-II)

In the watershed areas of Jammu and Kashmir, the irrigation system consists of local systems such as medium and small water harvesting structures and distribution channels constructed along the slopes of hills (locally known as *khuls)*. Over the period, due to neglect and mismanagement, their utilities in irrigating crops have declined remarkably. IWDP (Hills-II) has started the process of modernization and repair of the existing water harvesting structures and gravity based irrigation water distribution channels. Besides, treatment of the catchment areas has also been initiated to check erosion, restore the vegetative cover in the upper reaches, and remove impediments to the free flow of water into the water harvesting structures. Together, in course of time, these measures are also expected to increase the proportion of rainfall that is harnessed for local irrigation. User groups have been involved in repair, maintenance and improvement of the physical structures, as well as water management on cost-sharing basis mainly in the form of voluntary free labour. The project has introduced significant institutional reforms to revive and strengthens community institutions to make them more effective and flexible. It has played a vital role in clearly defining the relative rights and obligation for users communities, which are part of a larger integrated system as well as the rights of user communities with those outside. The attention has also been given to the basis on which entitlements can be changed, and putting in place an institutional mechanism for enforcing regulations. The project has recognized that better use of available infrastructure with user participation is probably more cost effective than creating new systems. As a result, improvement of the existing physical structures received high priority than the construction of new structures. For this, the project has allocated substantial resources.

(a) Formation of WUA

WUA is the basic foundation of the irrigation reform process under the IWDP (Hills-II). The process of formation of a WUA was relatively simple. The election process to form the WUAs was highly democratic and fair and was conducted in a participatory manner. Initially, WUAs have been formed in those areas where physical infrastructure for irrigation already existed in the form of water harvesting structures and small canal outlets in upper hill slopes fed by rains, which were in need of repairs and maintenance. The participatory development staff of the project along with village *panchayat* identified the user-farmers. The elections were held for the formation of WUAs, with active support of the *panchayat* and project functionaries. On the specified date, elections were conducted. Each voter casts his vote. The candidates securing the maximum numbers of votes were declared elected. In some cases, the elections may be unanimous. The consensus candidate was then declared elected. The tenure of the WUA is for a period of one year. When a vacancy arises due to death or resignation of a WUA president or a member, the vacancy is filled up following the similar process. Each WUA has a separate bank account. The president and a member are the signatories to the bank account.

(b) Operation of Irrigation System

To begin with, the entire reform process has been repair and renovation oriented. The majority of the works have been done through WUAs along with *panchayats*. WUAs have executed the works in a fair and impartial manner, with proper records of all financial transactions maintained, and contracts given in a fair and just procedure. In general, the cost of the works done by WUAs is 25 per cent lower than those carried out by the contractors. Most of the works are done at the estimated rate. A part of the sanctioned amount was paid in advance and remaining on the completion of the works. But in majority of the cases, even after the completion of the works, the balance amount was not paid. Whatever funds were released go directly to the WUA account. The project officials have no role in handling the funds. Proximity to the process of technical and financial estimates, funds disbursements, works execution by the local farmers, have all brought more transparency in WUA functioning. Contractors were almost out of the picture. The project authorities have given clear instructions that there will be no new structures, or alteration of original designs. So, the major emphasis is on restoring the original designs. The repairs and strengthening of structures, adding of missing structures, desilting the structures, installation of water regulation structures wherever damaged or not installed earlier, and strengthening the distribution channels and bunds are some of the major works carried out by the WUAs. In the process of users identifying and executing the works in case of drainage system, a lot of attention has been paid on users' needs like cattle pathways, bathing steps for women, removal of encroachments and clearing shrubs and silt for smooth flow of water till the tail-reach. Likewise in the case of water-harvesting structures, in addition to the above-mentioned works, the major focus was on repairing and strengthening the existing infrastructures, raising and strengthening bunds to original designs, and improving the water-harvesting structures through voluntary labour contribution. In some cases, the president of WUAs have been trapped in difficulties owing to lack of proper manpower and material planning, which resulted in severe losses to them.

(c) Allocation Rules

The operation aspects are necessary to ensure that physical facilities are maintained in good working condition and regulate access to facilities provided by it. One of the crucial factors on which the authority and credibility of the WUA depends is its ability to ensure that equitable distribution of the benefits takes place and any free rider is penalized. Towards this end several rules have been formulated. One's share of water is determined by the land acreage. This ensures an equitable distribution of the common resources and the prevention of the free riding. In addition, members also patrol together in order to avoid any breaching of the channel. The enforcement of the rules and making sure that whatever penalties are imposed are paid is an arena for the WUA to prove its authority and impartiality. In the project area, irrigated water is used essentially for paddy cultivation. The extent of area to be cultivated as well as the timing of the start of irrigation in a particular season is generally decided by the users' association as a whole in the light of available water supply and rainfall at the beginning of the season. As the entire service area cannot be irrigated, the system is confronted with the problem of rationing supplies. The mechanisms have been evolved to ensure equitable sharing of the shortfall in supplies among all users. Acreage rationing is but one element of the water rationing system. In times of shortage, besides rationing acreage, the water allocation is regulated even more strictly. It is ensured that all users share the reduction in both area and water supply in more or less the same proportion. As paddy is relatively water intensive crop grown in the project area, a system of rotational use of water is adopted even in a normal year to permit all users to grow the crop. The collective decisions are being taken at the beginning of the season regarding which land and how much is to be cultivated, which implies certain coordination in the timing of operations.

(d) Maintenance of Irrigation System

The purpose of 'maintenance' is to ensure that physical facilities function smoothly and at the level of performance for which they were designed. If the maintenance is inefficient, the volume of water made available to the fields get reduced, which hampers the level of output in relation to the potential of the system. The quality of the maintenance affects the interests of both the organizations and the users. The strength of this common interest is stronger in PIM, as the users have made a substantial contribution to the cost of developing it. The common interest of user organizations is also strong, where the neglect of maintenance leads to heavy loss in productivity. WUAs have established conventions regarding the timing of repairs, division of work, responsibility of members and the obligations of users. The custom and social pressure facilitates the smooth working of the arrangements. However, they do not work entirely on the strength of mutual interest reinforce by custom. There is provision of sanction (ranging from fines to loss of water rights) against non-compliance, for which WUA has the authority to enforce the sanction to ensure compliance. However, not even a single case has been reported of sanction against non-compliance. The large landowners have dominated the WUA functioning, as they cultivate a large proportion of land and benefits relatively more from the irrigation system. Thus, the land tenure system has very closely influenced the management of local water control systems.

(e) Benefits and Costs

The potential beneficiaries were mobilized on issue related to fair sharing of the benefits and costs. It was emphasized that users' participation will result in cost-effectiveness in repairs and maintenance and efficient management once it is commissioned. Since a part of the cost of resource development and operation mainly in the form of labour contribution is to be borne by the potential beneficiaries, they have keen interest in minimizing the costs, ensuring a fair balancing of competitive interests and providing economical and efficient services. There was provision of equitable distribution of water based on the size of land holdings in the area of their operation and this has been adhered to in actual practice also. However, there was an unscrupulous distribution here and there, and conflicts occur among the users and members of users' association over the way allocations are managed and over attempts to violate the allocations. In such situations, the executive committees of the WUA mediate disputes among its members. Disputes that cannot be resolved at WUA level are referred to village *panchayat*. The conflicts, which cannot be resolved by the WUA as well as *panchayat*, are referred to project officials. It does not however follow that the latter will always intervene decisively.

(f) Transparency and Accountability

Transparency is one of the key principles of PIM. In order to strengthen confidence over the functioning of the users' organizations, the activities undertaken by them have to be informed to all the water users in the area of their operation. PRA exercises reveal that farmers have been informed in advance by the WUAs about the various activities to be undertaken. The management committees of the WUAs were disseminating information on various activities undertaken for efficient management of the irrigation systems in the area of their operation. The dissemination of information was through general body meetings on regular intervals. The steps have been undertaken to prioritize the repair works on the system so that the available funds should be utilized precisely. Keeping in view the limited amount of funds available at their disposal, the prioritization of the work was also considered necessary by the majority of the users. The quality of the works undertaken by the WUAs were comparatively better than the quality of the works undertaken by the contractors prior to the system of PIM. The conveyance losses in the distribution channels have been minimized, which facilitated in distribution of water up to the tail end area. With PIM, the project functionaries are playing the role of facilitators and made accountable and responsible to the farmers' organizations. On the other hand, the executives of the WUAs are also accountable to their members. The provision of re-election of the WUAs at regular interval also ensured the downward accountability in the reform process.

IMPACT OF PIM

There are perceptible economic gains to the farmers after the initiation of PIM. The land values have increased in both forested as well as agricultural sub-watersheds in the range of 10-15 per cent. The irrigated land value is marginally more in agricultural sub-watersheds due to the field drainage provided by the WUAs, which also arrested the waterlogged conditions in the area of operation of the WUAs. The formation and operationalization of the WUAs have resulted in an increase in area under cultivation and ensure timely sowing of crops and receipt of water at the tail end. New acreage has been added to irrigate lands due to better coverage and maintenance. As a result, irrigated area has increased mainly in tail-reaches. With better maintenance of the irrigation system, the availability of water for various crops has increased, which resulted in increased productivity in terms of rice, pulses, vegetables and orchards. The level of income derived from various crops has increased significantly as compared to the income levels prior to the formation of WUAs. The rate of increase in income has been reported to be comparatively higher in rice and vegetables than other crops. The area of operation of WUAs was under the rice-growing belt, and the farmers reported that the cropping pattern may conveniently be shifted to some extent to commercial crops like vegetables and pulses, which may assure higher returns in future.

(a) Reduction in Rainwater Loss and Sediment Yield

The micro-level watershed planning was carried out using the sweeping transect and emphasis has been given on soil erosion control on hill slopes and gullies, regulation of water flow system in the watershed drainage, and rearrangement of farmlands. The diversion channels were made along all the hills and hillocks to guide the high velocity runoff from these sites on a safe and controlled way, so that the runoff with soil loads should not enter the arable land and cause further degradation. The adverse climatic factors of the *Shivaliks* necessitated the adoption of micro catchment techniques for run-off harvesting and conservation practices as done in similar areas elsewhere. The usefulness and scope of rainwater harvesting and conservation practices in improving tree growth in arid zones have been amply demonstrated. The staggered trenches ensured higher survival rate of out planted seedlings in the experimental plots. The contour trenches also helped in moisture retention but the most of the sub-watershed of Akhnoor being an undulated terrain without sufficient slopes–the contour interval had to be very wide. The effect of regeneration of vegetation, along with soil and water conservation measures on hillslopes and wastelands was substantial. In *Shivaliks*, the run-off soil loss on barren hills was 23. 70 tons in the baseline period, which fell to 9.65 tons per hectare after treatment. In *Karewas*, it declined to 4.80 tons per hectare from 8.24 tons per hectare in baseline period (IWDP, 2001). The progressive reduction in soil loss and sediment yield as a result of quick recovery of vegetation on hillslopes and lands adjoining the foothills has resulted in a greatly improving surface and ground water regime of the selected sub-watersheds.

(b) Status of Water Resources and Irrigation

The table 1 gives the data on surface and ground water resources across the sampled sub-watersheds. Both the average number of water points (*bowlies*) as well as the gravity based water points per village were comparatively more in project area with WUA than without WUA and non-project area. In the project area with WUA, there were 6.8 and 5 natural water points and gravity based water points respectively in forested watershed and 4.3 and 2 respectively in agricultural watershed. In non-project area, the average number of natural and gravity based water points

Table 1
Status of Water Resources and Irrigation (per village)

Item	*Unit*	*Project area with WUA*		*Project area without WUA*		*Non-project area*	
		Forested WS	*Agrl. WS*	*Forested WS*	*Agrl. WS*	*Forested WS*	*Agrl. WS*
Surface Water Supply							
Natural Water Points (Bowlies)	No.	6.8	4.3	5.2	3.4	3.6	2.6
Gravity based Water Points	"	5	2	2	1	1	0
Ground Water Supply							
Hand Pumps	"	2	3	0	1	0	1
Water Harvesting Structures	"	3.5	2.9	2	1.8	1.2	0.7
Irrigation							
Irrigation Channel	No.	2.8	1.7	1.8	1.2	1	0.8
Gross Irrigated Land	Ha.	98	67	48	64	69	44
Net Irrigated Land	Ha.	68	36	**34**	37	27	18
Irrigation Intensity	%	144.11	186.11	141.17	172.97	255.56	244.44

Note: Irrigation Intensity =Gross Irrigated Area / Net irrigated Area x 100.

were comparatively very low and stood respectively at 1.6 and 1 in forested watershed, and 2.6 natural water points in agricultural watershed, whereas there was no gravity based water points in agricultural watershed. This clearly reflects the benefits of PIM through which the project has implemented the repairs of water resources points and installed new water points, mainly gravity based water resources. In the forested watershed, the difference between project area with WUA and non-project area in average number of natural water points and gravity based water points was as high as 3.2 and 4 respectively. The average number of hand pumps per village was also reportedly high in project area with WUA than project area without WUA and non-project area.

A mere perusal of the table 1 makes it clear that the number of water harvesting structures were more in project area with WUA than project area without WUA and non-project area. Similar is the case with the length of irrigation channel. However, significant differentials were noticed across project area with WUA and without WUA as well as non-project area (see table 2). The number of water harvesting structures as well as length of irrigation channel was significantly higher in project area than non-project area. In the project area with WUA, the average number of water harvesting structures per village was 3.5 and 2.9 respectively in forested and agricultural watersheds, whereas, it was 2 and 1.8 respectively in project area without WUA. In non-project area, the average number of water harvesting structures was 1.2 and 0.7 respectively in forested and agricultural watersheds. It is significant to note that the quality of water harvesting structures were also reportedly better in project area with WUA than non-project area, due to participatory repairs and renovation activities as well as maintenance initiated by the project. Similarly, the average length of irrigation channel in the project area with WUA was 2.8 km and 1.7 km per village respectively in forested and agricultural watershed, whereas, it was 1.8 km and 1.0 km respectively in project area without WUA. In the non-project area, the average length of irrigation channel was comparatively low and it was just 1.0 km and 0.8 km per village respectively in forested and agricultural watersheds. As a result of increased water potential in the selected sub-watersheds, the area under irrigated cropping increased significantly.

Table 2
Difference in Status of Water Resources and Irrigation in Sample Area (per village)

Item	*Unit*	*Project area with WUA*		*Project area without WUA*		*Non-project area*	
		Project area without WUA		*Non-project area*		*Non-project area*	
		Forested WS	*Agrl. WS*	*Forested WS*	*Agrl. WS*	*Forested WS*	*Agrl. WS*
Surface Water Supply							
Natural Water Points (Bowlies)	No.	1.6	0.9	3.2	1.7	1.6	0.8
Gravity based Water Points	"	3	1	4	2	1	1
Ground Water Supply							
Hand Pumps	No.	2	2	2	2	0	0
Water Harvesting Structures	"	1.5	1.1	2.3	2.2	0.8	1.1
Irrigation							
Irrigation Channel	No.	1	0.5	1.8	0.9	0.8	0.4
Gross Irrigated Land	Ha.	50	3	29	23	-21	-20
Net Irrigated Land	Ha.	34	-1	41	36	7	19
Irrigation Intensity	%	2.94	13.14	-111.45	-58.33	-114.39	-71.47

Since changes in irrigated area depend on numerous factors, many of which could not be measured in the current study, the findings here must be treated cautiously. The data presented in table 1 reveals that the changes in gross and net irrigated area in project area with WUA and without WUA, and non-project area. A perusal of the table makes it clear that irrigation intensity in forested and agricultural sub-watersheds in project area with WUA was 144.11 per cent and 186.11 per cent respectively. In project area without WUA, it was 141.17 per cent and 172.97 per cent, whereas in non-project area, it was 255.66 per cent and 244.44 per cent respectively in forested and agricultural sub-watersheds. The data shows the marked differences in irrigation intensity across the sample area. The difference in irrigation intensity was 2.94 per cent and 13.14 per cent respectively in forested and agricultural sub-watersheds of project area with WUA and without WUA. It is significant to note that the differences in irrigation intensity in project area with WUA and non-project area (see table 2).

(c) Change in Crop Intensity and Crop Productivity

As a result of the improved soil moisture regime, the increase in supplementally irrigation resources and increased use of fertilizer (including cow-dung), the cropping intensity has improved in the project area compared to non-project area. The data on grossed crop area, net area sown and crop intensity is presented in table 3, which clearly reveals significant variations across the selected sub-watersheds. There are considerable differences in cropping intensity in project area with WUA and non-project area, which can be attributed to the project interventions in the field of agriculture through rainfed crop demonstration, propagation of the use of modern inputs through extension agents of the project. It is to be noted that differences in cropping intensity was reportedly negative across the project area with WUA and without WUA as well as non-project area. For instance, it was–58.26 per cent and–46.27 per cent respectively between project area with WUA and non-project area (see table 4).

It is very difficult to isolate the impact of various inputs used on crop yields. Due to creation of VCLs and forest closures, an increase in the vegetation has taken place. The over all soil-moisture regimes have also improved after project intervention. The project has repaired the traditional irrigation water distribution channels (khuls), consequently the irrigation water availability and flow has improved. Keeping the influence of these factors aside, the data presented in table 4 shows differential levels of yields of various crops such as maize, paddy, pulses and wheat across the selected sub-watersheds, largely due to increased use of non-seed inputs. A cursory glance at table 4 makes it evident that average crop yield per hectare has been comparatively more in project area with WUA than project area without WUA, and further the yield per hectare has been comparatively higher in project area without WUA than non-project area. The productivity of selected crops is comparatively higher in agricultural sub-watersheds than forested sub-watersheds. However, a significant variation in crop yield is reported across the project and non-project area.

The yield difference in the project area with WUA and without WUA was low compared with project area with WUA and non-project area in forested as well as agricultural sub-watersheds. The yield difference between project area with WUA and non-project area was 2.1 quintals, 0.8 quintal, 0.64 quintal and 1.36 quintal per hectare respectively for maize, paddy, pulses, and wheat in forested watershed. Whereas, in agricultural watershed, yield differences between project area with WUA and non-project area was markedly high and stood at 2.2 quintals, 1.7 quintals, 0.5 quintal, and 2.1 quintals respectively for maize, paddy, pulses, and wheat. However, such differentials were not so marked in project area without WUA and non-project area. The yield differences were 0.8 quintal, 0.1

quintal, 0.1 quintal and 0.6 quintal respectively for maize, paddy, pulses and wheat in agricultural watershed, and more or less the same is true for forested watershed (see table 4).

Table 3
Crop Intensity and Crop Yields

Item	*Unit*	*Project area with WUA*		*Project area without WUA*		*Non-project area*	
		Forested WS	*Agrl. WS*	*Forested WS*	*Agrl. WS*	*Forested WS*	*Agrl. WS*
Gross Cropped Area	Ha.	163	74	73	107	113	68
Net Sown Area	Ha.	158	68	57	93	70	42
Crop Intensity	%	103.16	115.63	128.07	146.58	161.42	161.9
Non-seed Inputs and Crop Yields							
Maize	Qntl./Ha	96	9.9	8.2	8.5	7.5	7.6
Paddy	"	5.2	6.2	4.5	8.5	4.4	4.5
Pulses	"	2.64	2.8	2.3	2.4	2	0.3
Wheat	"	8.96	9.9	8.2	8.4	7.6	7.8
Seed Inputs and Crop Yields							
Maize							
Local	Qntl./Ha	13.13	14.2	12.6	13.7	10.3	11.2
Hybrid	"	15.6	15.8	14.9	14.1	11.7	11.9
Difference	"	2.3	1.6	2.3	0.4	1.4	0.7
Paddy							
Local	"	10.8	11.2	9.4	10.3	9	9.2
Hybrid	"	12.3	13.1	11.4	11.7	10.1	10.7
Difference	"	1.5	1.9	2	1.4	1.1	1.5
Wheat							
Local	"	18.4	19.7	16.9	17.4	16.1	16.7
Hybrid	"	21.3	21.8	20.4	21.1	17.3	17.9
Difference	"	2.9	2.1	3.5	3.7	1.2	1.2

Note: Crop Intensity= Gross Cropped Area / Net Area Sown x 100.

The data presented in table 4 further substantiate the yield differentials across the selected sub-watersheds. As can be noticed elsewhere too, there are significant differences in yield of local and hybrid varieties of crops in forested as well as agricultural sub-watersheds. For instance, yield difference in local and hybrid maize is 2.3 quintal per hectare in forested sub-watershed in project area with WUA and without WUA. Thus, one can infer that new institutional arrangements have no impact on the yield of local and hybrid variety. The yield difference is 1.2 quintal per hectare in agricultural sub-watershed in project area with WUA and without WUA. A significant differential in yield is noticed between project area with WUA and non-project area, as well as project area without WUA and non-project area. More or less the same applied to other crops with minor differences across the forested and agricultural sub-watersheds. On the whole, the average crop yield per hectare has significantly improved after the project area.

Table 4
Difference in Crop Intensity and Crop Yields in Sample Area

Item	*Unit*	*Project area with WUA*		*Project area without WUA*		*Non-project area*	
		Project area without WUA		*Non-project area*		*Non-project area*	
		Forested WS	*Agrl. WS*	*Forested WS*	*Agrl. WS*	*Forested WS*	*Agrl. WS*
Gross Cropped Area	Ha.	90	-33	50	6	-40	39
Net Sown Area	Ha.	101	-25	88	26	-20	51
Crop Intensity	%	-24.91	-30.95	-58.26	-46.27	-33.35	-15.32
Non-seed Inputs and Crop Yields							
Maize	Qntl./Ha	1.4	1.4	2.1	2.2	0.7	0.8
Paddy	”	0.7	1.6	0.8	1.7	0.1	0.1
Pulses	”	0.34	0.4	0.64	0.5	0.3	0.1
Wheat	”	0.76	1.5	1.36	2.1	0.6	0.6
Seed Inputs and Crop Yields							
Maize							
Local	Qntl./Ha	0.7	0.5	3	3	2.3	2.5
Hybrid	”	0.7	1.7	3.9	3.9	3.2	2.2
Difference	”	0	1.2	0.9	0.9	0.9	-0.3
Paddy							
Local	”	1.4	0.9	1.8	2	0.4	1.1
Hybrid	”	0.9	1.4	2.2	2.4	1.3	1
Difference	”	0.5	0.5	0.4	0.4	0.9	-0.1
Wheat							
Local	”	1.5	2.3	2.3	3	0.8	0.7
Hybrid	”	0.9	0.7	4	3.9	3.1	3.2
Difference	”	-0.6	1.6	1.7	0.9	2.3	2.5

Besides above, the repairs and maintenance works have generated much needed wage employment for the poor labourers and small and marginal farmers. The undertaking of the construction works by the members of WUAs along with *panchayats* has resulted into cost-effectiveness to the tune of about 10-15 per cent. PIM have also empowered the farming communities in self-organization and equipped them with the supervision and management skills. The farmers' participation in various stages of the reform process has created a sense of ownership of the assets created. They have played a greater role in the choice of prioritizing and executing the work themselves through the own organizations. Above all, the institutional changes facilitated the closer interaction of the project functionaries and local stakeholders and unified the diverse groups around the water.

CONCLUSION

The institutional reforms in irrigation water management initiated under IWDP (Hills-II) are both bold and innovative. The WUAs have taken up the maintenance activity with the financial support provided by the project. Due to physical improvements in the irrigation system, the water availability has improved. It addresses the key issues pertaining to irrigation management: institutional structure, incentives, accountability, transparency, and sustainability. The institutional changes are still in progress. In the future, linkages need to be established which will make WUAs independent of project support. There is need to redefine the role of irrigation agency, which calls for suitable institutional restructuring. Ultimately, WUAs will need to be financially self-sustainable. The sustainability of the WUAs essentially depends on their capacity to operate and manage the system on their own. The financial sustainability has an important role in the sustainability of the WUAs. At present, WUA is completely dependent on the project functionaries for funds. The physical works will continue for some time, which would provide some finances for the day-to-day functioning of the WUA. But what happens when the funds from the donor stop. In the past, in other parts of the country where WUAs were formed, many became defunct due to financial bankruptcy. WUA will have to be financially self-sufficient to perform their duties on sustainable basis and to make the reform process a success. In the near future, it is envisaged that the WUAs will collect water charges from the users. This will be the most logical step in the reform process. However, the WUA executives may not be able to perform their duties in this regard due to inadequate manpower at their disposal. In order to operationalize the task of water tax collection, concrete steps need to be taken. The steps need to be initiated to strengthen the capacity building of WUAs to ensure financial sustainability, after the withdrawal of the project.

Another weak link has been the issue of accountability. Mechanisms to ensure accountability must continue to be evolved. The roles and responsibilities of all the agencies concerned must be further defined. There is need to transfer all the operation and maintenance function to the WUAs in near future. Likewise, transparency is critical if the organizations are to progress further. To a large extent, transactions are captured in the accounting system. There is need to put in place simple and standardized procedures for accounting and finance. There is need to ensure replicability of the successes achieved. In the tasks of operation and maintenance of the PIM, emphasis should be paid on getting rid of the middlemen and contractors. Users' participation should be ensured for identifying, and executing the essential works for repair and maintenance. The role of the project functionaries or line department needs to be curtailed. At the same time, there is need to form federations of WUAs, for which continued support and training are required. The sustainability of the farmers' organization also calls for involvement of NGOs and training institutes in upgrading the skill and capacity building. There is urgent need to forge suitable linkages with reputed local NGOs, training institutes and line departments to equip the farmers' organization in various aspects of PIM. No doubt, adequate groundwork and the required environment are created, there is need to give a closer look to the weaknesses and remedy them. More importantly, a regular monitoring mechanism has to be put in operation to initiate corrective measures as and when needed.

REFERENCES

Bandyopadhyay, J. and S. Perveen (2004), Interlinking of Rivers in India: Assessing the Justifications, *Economic and Political Weekly*, 39 (50): 5307-16.

Bandyopadhyay, J., B. Malik, M. Mandal and S. Perveen (2002), Dams and Development Report on a Policy Dialogue, *Economic and Political Weekly*, 37 (4): 4108-12.

Brewer, J., S. Kolavalli et al. (1999), *Irrigation Management Transfer in India: Policies, Processes and Performance*, IIM, Ahmedabad and IIMI, Colombo, Oxford and IBH Publishers.

Deursen, W.P.A. (2000), *Humans, Water and Climate Change*, The Dutch National Research Programme on Global Air Pollution and Climate Change, Carthago Consultancy, Rotterdam.

Falkenmark, M. et al. (1980), *Water and Society: Conflicts in Development*, Part 2, Pergamon Press, Oxford.

Gleick, P.H. (1996), Basic Water Requirement for Human Activities: Meeting Basic Needs, *Water International*, 21: 83-92.

IWDP (2001), *Status Report for Supervision Mission of World Bank*, IWDP (Hills-II), Jammu and Kashmir.

Ives, J.D. and B. Messerli (1989), *The Himalayan Dilemma*, Routledge, London.

IWRS (1996), Interbasin Transfers of Water for National Development: Problems and Perspectives, Indian Water Resources Society.

Iyer, R.R. (2000), *Water: Charting a Course for the Future*, Centre for Policy Research, New Delhi.

Khattri, K.N. (1987), Great Earthquake Seismicity Gaps and Potential for Earthquake Disaster Along the Himalayan Plate Boundary, Techno-physics, Vol. 138, pp. 79-82.

Nigam, A., B. Gujja, J. Bandyopadhyay, and R. Talbot (1997), *Freshwater for India's Children and Nature*, WWF and UNICEF, New Delhi.

Parthasarathy, R. (2000), Participatory Irrigation Management Programme in Gujarat: Institutional and Financial Issues, *Economic and Political Weekly*, 25 (35 and 36): 3147-54.

Serageldin, I. (1995), *Towards Sustainable Management of Water Resources*, World Bank, Washington, D.C.

Vidal, J. (2003), India's Dream Bangladesh's Disaster, *The Guardian*, July 24.

World Commission on Dams (2000), *Dams and Development–A New Framework for Decision-Making*, Earthscan Publishing, London.

34

DRYLAND BASED FARMING SYSTEM MODELS FOR SUSTAINABLE AGRICULTURAL DEVELOPMENT

G.N. Nagaraja, Rajeswari K. and Sukanya C.

Agricultural sector continues to be the pivotal sector in the Indian economy providing employment to 69 per cent of the population but its contribution to the Gross Domestic Product (GDP) is only about 23 per cent. Out of an estimated 143 million ha net cultivated area in India, about 97 million ha. (68%) is dryland producing 44 per cent of the country's food requirements and supporting 40 per cent of human and 60 per cent of livestock population (NBSSLUP, 2001).

Even when the full irrigation potential of 139.5 million ha is realized by 2050 in agriculture, 75 million ha will continue to be solely dependent upon rainfall. About 15 million ha of dryland lies in the arid region and receives less than 500 mm rainfall, another 15 million ha is in 500-750 mm rainfall zone, 42 million ha. is in 750-1150 mm rainfall zone and remaining 25 million ha receives more than 1150 mm rainfall per annum (Kanwar, 1999).

Of the 97 million farm holders, 76 per cent are small (< 2 ha) and marginal, cultivating only 29 per cent of the consolidated and scattered arable land. Hence, farmers' dependence on livestock besides arable farming, as an alternative source of income is very high. Preliminary estimates revealed that nearly two out of three heads of cattle population of 219 million thrive in the Indian drylands (FAO, 2000). Further, resource poor farmers, poor infrastructure and low investment in technology input characterize the Indian drylands.

Land being the most limited and scarce resource, particularly on small and marginal farms, the scope to increase farm income and employment through crop-production alone is not too bright. Therefore, one has to look for alternatives in order to get assured increase in the employment of the weaker sections. In this regard integrated farming system is the only answer in which livestock, poultry, piggery and other allied activities are regarded as the important components. These components on farms would not only reduce the magnitude of seasonal variability in unemployment but also become a source of continuous income throughout the year.

The total area under cultivation in Karnataka is about 102.58 lakh ha of which, nearly 80 per cent of the cultivated land falls under rainfed farming. Out of the 10 agro-climatic zones, 5 zones can be classified broadly as drought prone areas, which includes eastern dry zone of Karnataka. The eastern dry zone covers Kolar (11 taluks), Bangalore rural (9 taluks) and Tumkur (2 taluks) districts in Karnataka state. In the present paper, an attempt is made for identification and economic appraisal of the various dryland based farming systems in Karnataka in general and those of the Kolar district in particular (Rajeswari, 2003).

METHODOLOGY

The total geographical area in the Kolar district is 7,79,467ha, of which Bagepalli and Srinivasapur taluks accounts for 90,009 ha and 85,824 ha, respectively (2001-02). The net area under un-irrigated land, which reflects the total dryland area of the district, is 2,82,504 ha of which Bagepalli and Srinivasapur taluks accounts for 25,717 ha (9.1%) and 33,791ha (11.96%), respectively.

Kolar district was purposively selected for the study, since 77.92 per cent of the total net cultivated area (3.6 lakh ha.) is under dryland where farmers are following different types of farming systems. Again two taluks were selected for the study since 79 per cent and 92.6 per cent of the area under cultivation is rainfed in Bagepalli and Srinivasapur taluks, respectively. The main objective of the investigation is to identify, assess and design appropriate dryland based farming system models for marginal and small farmers who form 57 per cent and 20 per cent of the total size of holdings and area under cultivation in the district.

This study has a special focus on dryland based farming systems. In the first stage, two taluks were selected out of eleven taluks in Kolar district based on the area under dryland. In the second stage, six villages were randomly selected in probability proportion to total number of villages in each taluk based on area under dryland. A sample of 86 farmers having only rainfed lands were selected through simple random sampling method from among the identified farming systems within the selected villages such that each enterprise generated at least 10 per cent of the total farm income. The data were collected from the selected farmers during the month of July and August 2003. The information collected related to only Kharif season of 2002-03 on cropping system and their allied enterprises like dairy, piggery, sheep, and horticulture.

LINEAR PROGRAMMING

Linear programming technique formed the basis for obtaining the optimal solution for different enterprises followed by the sample farmers. Optimum allocation of resources was defined as one, which, given physical, technical and resource conditions shows the activities to undertake and how much

of each resource to allocate to each activity so that the net farm returns are maximized in a year (Nagaraja, 1995).

Mathematical formulation of the model

Maximize $Z = \sum_{j=1}^{n} C_j X_j$

Subject to the constraints,

1. $\sum_{j=1}^{n} a_{ij} X_j \geq b_i \quad (j = 1 \ldots\ldots k)$

2. $\sum_{j=1}^{n} a_{ij} X_j \geq bi \quad (j = k+1 \ldots\ldots\ldots m)$

3. $\sum_{j=1}^{n} a_{ij} X_j = b_i \ (j = m+1 \ldots\ldots\ldots n)$

4. $X_j \geq 0$

Where,

Z = Net returns from all crop and allied activities included in the model.

C_j = Net returns from j^{th} activity measured in rupees per hectare of land.

X_j = Level of jth activity in hectare (for crop/horticulture activity) and in unit (for allied activities).

a_{ij} = Quantity of ith input required per unit of jth activity.

b_i = Quantity available of the ith resource.

Objective Function

The objective function for the model in this study is the maximization of the annual net returns to owned resources. The gross returns per hectare of crop and per unit of allied activities were calculated by using the data of sample farmers. Paid out costs such as hired human labour, bullock labour, fertilizer etc. were directly subtracted from the gross returns. For dairying, sheep and poultry activities, gross profits are calculated by deducting variable expenses such as value of fodder, concentrates, animal health and veterinary care from gross income of dairy, sheep and poultry activity. The maximization of net returns (profit) is subject to the assumption and resource constraints imposed in the model. In the case of horticulture activity, the annual operating expenses of maintenance of horticulture (mango) garden was added to the annual production costs of mango production and then deducted from gross return to get gross margin.

Constraints and Requirements

Some of the important constraints considered in the study are land, labour, consumption, fodder availability and borrowing.

Activities in the model: Activities specify the resources, which could be put into various alternative uses. The activities included in the model are:

1. Crop, dairy, poultry, sheep and mango production activities.
2. Labour hiring activities.
3. Product and labour sale activities.
4. Borrowing activities.
5. Consumption activity.

Input-output co-efficient

The input co-efficient in this study pertained to land, labour, capital, FYM, fodder availability, fertilizer and fodder requirement. Input co-efficient is the average requirement of the ith input per unit of jth activity. It is calculated on per acre basis in case of crop and horticulture enterprises. The input co-efficients for different farm categories derived by totalling and averaging the relevant input-output data. For dairy enterprise it is on per animal basis. To accomplish the objective of maximization of gross margins to the family owned resource, the basic linear programming model is employed in this study (Nagaraja, 1995).

RESULTS AND DISCUSSION

Existing cropping patterns, income and employment generated from farming systems models

In the case of marginal farmers, the average land holding size was 0.68 hectares, whereas in the case of small farmers, it was 1.47 hectares. The cropping system followed by different categories of farmers in the area under study was dryland based: ragi and groundnut with mixed cropping (intercrops like red gram, field bean, jowar), fodder crops (jowar and maize) and horse gram formed the major crops. In this situation, most of the family food grain consumption needs and fodder requirement for the livestock were met by own farm production. The priority in cropping pattern decisions by the farmers is because the crops are very much suited to rainfed conditions.

Marginal farmers cultivated groundnut with intercrops in 0.32 hectares, ragi with intercrops in 0.13 hectares, fodder maize in 0.03 hectares, horse gram in 0.07 hectares and mango in 0.14 hectares. Whereas small farmers cultivated 0.24 hectares of groundnut with intercrops, 0.23 hectares of ragi with intercrops, 0.29 hectares of fodder crops, 0.06hectares of horse gram and 0.66 hectares of mango, respectively (See table 1).

The sample farmers also cultivated dryland horticulture crop mango, which is a rainfed crop. In the existing cropping systems, dryland horticulture occupied more proportion of area in the total size of land holdings in the case of small farmers, whereas groundnut with mixed cropping followed by ragi (with intercrops) and mango occupied the major area in marginal farmers' category.

Table 1
Existing Cropping Pattern of Sample Farmers

(in hectares)

Particulars	*Marginal farmers*	*Small farmers*
Groundnut + red gram	0.12	0.10
Groundnut + red gram + field bean	0.20	0.14
Ragi+ field bean	0.05	0.14
Ragi + red gram	0.08	0.09
Pure fodder maize	0.03	0.07
Pure fodder jowar	-	0.22
Horse gram	0.07	0.06
Mango *	0.14	0.66
Total	0.69	1.48

Note: *Ragi is grown as an intercrop in mango orchard upto 4th year of Establishment.

Farming system is aimed at the efficient use of resources to maximize the income. It also minimizes the production risk by spreading the risk to various enterprises instead of one activity. The details of net farm income derived from the existing farming system are furnished in the table 2. It was found that the marginal farmers realized highest net income of Rs. 31,000 from C+D+H+Sh system followed by Rs. 27,210 from C+D+H system. With respect to small farmers, C+D+H+Sh gave a highest net farm income of Rs. 49,250 followed by C+D+H Rs. 42,800, respectively.

Table 2
Net farm Income from the Existing Farming Systems among Marginal and small farmers

(in rupees)

Farming systems	*Marginal farmers*	*Small farmers*
C+Sh	6,250	9,000
C+D	10,400	17,125
C+H	19,000	-
C+D+Pi	20,000	-
C+D+Sh	12,500	20,667
C+D+H	27,210	42,800
C+D+H+Sh	31,000	49,250

Note: C=Crop, D=Dairy, Sh=Sheep, H=Dryland Horticulture, Pi=Piggery.

Marginal and small categories of farmers derived highest farm income from C+D+H+Sh system followed by C+D+H. The main reason behind this is that these systems involve dryland horticulture enterprise and dairy enterprise which give high income to the farmers compared to the other enterprises. Hence, it can be concluded that crop, dairy and dryland horticulture formed the major income generating components of the farming systems in the area under study.

Apart from the income, the various farming systems generated human and animal labour employment among the two categories of farmers. The details on quantum of employment are presented in table 3. Under the existing farming systems followed by marginal farmers, C+Sh+D system provided highest employment of 319 mandays, 190 womandays and 17 bullockdays and in the case of small farmers, C+D+H+Sh system provided highest employment of 483 mandays, 481 womandays and 23 bullockdays.

Table 3
Labour Employment in the Existing Farming systems among Marginal and small farmers

(in labour days)

Categories	*Marginal farmers*			*Small farmers*		
Farming systems	*M*	*W*	*BP*	*M*	*W*	*BP*
C+Sh	240 (20)	215 (35)	11 (11)	209 (149)	120 (20)	9 (0)
C+D	131 (17)	160 (8)	8 (8)	197 (35)	131 (42)	14 (4)
C+H	75 (25)	59 (19)	12 (12)	-	-	-
C+Sh+D	319 (54)	190 (30)	17 (17)	320 (120)	235 (140)	10 (4)
C+D+Pi	193 (25)	174 (30)	12 (12)	-	-	-
C+D+H	212 (20)	123 (33)	10 (10)	376 (111)	317 (119)	19 (10)
C+D+H+Sh	300 (60)	220 (40)	15 (15)	483 (200)	481 (225)	23 (13)

Note: 1. C=Crop, D=Dairy, Sh=Sheep, H=Dryland Horticulture, Pi=Piggery.
2. M=Man, W=Woman, BP=Bullock pair.
3. Figures in the parentheses indicate the quantum of hired labour.

In addition to family labour, the farmers hired both human and bullock labour to perform various farm operations. It was found that under the existing farming system, marginal farmers hired 60 mandays, 40 womandays and 15 bullockdays per year in C+D+H+Sh system. Among small farmers' category, C+D+H+Sh system required 200 mandays, 225 womandays and 13 bullock days of hired labour.

In the case of small farmers, C+D+H+Sh and C+D+H systems generated more employment opportunities. As they could afford to pay the wages, they hired labourers to rear sheep of their farm. Hence, C+D+Sh system generated less employment to the family labour compared to the marginal farmers' category. This indicates the importance of livestock enterprise and its role in providing employment to both the categories of farmers.

NORMATIVE FARM PLANS MAXIMIZATION OF NET INCOME

Marginal Farmers

In order to optimize the objective of maximization of net income, the farmers need to adopt the combination of the enterprises as indicated in the normative plan presented in table 4.

The normative plans developed indicates an increase in the net farm income from Rs. 6,250 in the existing farming system to Rs. 8,250 (32%),

Rs. 10,400 to Rs. 13,416 (29%), Rs.19,000 to Rs.25,069 (31.94%), Rs.12,500 to Rs. 16,300 (30.40%), Rs. 20,000 to Rs. 27,560 (37.80%), Rs. 27,210 to Rs. 34,427 (26.52%) and Rs. 31,000 to Rs. 42,897 (38.38%) from models C+Sh (M_1), C+D (M_2), C+H (M_3), C+D+Sh (M_4), C+D+Pi (M_5), C+D+H (M_6) and C+D+H+Sh (M_7) respectively (see fig 1).

Table 4
Normative farm plans for Maximization of Net Income of Marginal farmers

(area in hectares)

Farming System Activity	*C+Sh M_1*	*C+DM_2*	*C+HM_3*	*C+D+Sh M_4*	*C+D+Pi M_5*	*C+D+H M_6*	*C+D+H+ ShM_7*
Groundnut + red gram (8:2)	0.25	0.20	0.16	0.25	0.29	-	-
Groundnut + red gram+ field bean (8:2:1)	0.08	-	-	-	-	-	-
Ragi + red gram (8:2)	0.35	0.34	0.32	0.24	0.25	0.22	0.12
Fodder maize		0.14	-	0.20	0.14	0.08	0.06
Mango	-	-	0.20	-	-	0.38	0.50
Dairy (in nos.)	-	1	-	1	1	1	1
Sheep (in nos.)	12	-	-	12	-	-	12
Piggery (in nos.)	-	-	-	-	4	-	-
Net income (in rupees)	8,250	13,416	25,069	16,300	27,560	34,427	42,897
Percentage increase in income over existing farming system	32.00	29.00	31.94	30.40	37.80	26.52	38.38

Note: C=Crop, D=Dairy, Sh=Sheep, H=Dryland Horticulture, Pi=Piggery

Figure 1
Comparison between Exisiting and Optimal net Income under Maximisation for Marginal Farmers

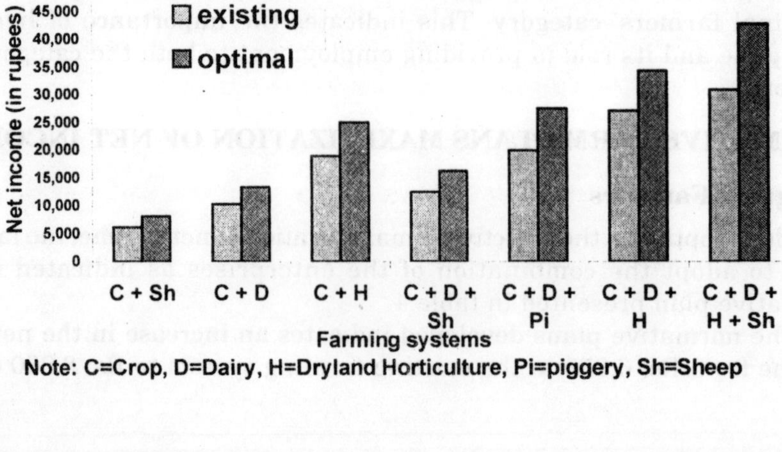

Note: C=Crop, D=Dairy, H=Dryland Horticulture, Pi=piggery, Sh=Sheep

The farming systems suggested that it requires cultivation of 0.25ha 0.20 ha 0.16 ha 0.25 ha and 0.29 ha of land under HYVs of groundnut + red gram with ratio of 8:2 proportion in models, M_1, M_2, M_3, M_4, and M_5, respectively. HYVs of ragi + red gram with the ratio of 8:2 in 0.35 ha, 0.34 ha, 0.32 ha, 0.24 ha, 0.25 ha, 0.22 ha and 0.12 ha in the models M_1 to M_7, respectively, helped to get highest net income. Fodder maize is suggested to cultivate in M_2, M_4, M_5, M_6 and M_7 models in 0.14 ha 0.20 ha, 0.14ha 0.08 ha and 0.06ha, respectively for maximization of income. In the case of dryland horticulture, mango has to be grown in an area of 0.20 ha, 0.38ha and 0.50 ha in M_3, M_6 and M_7 models, respectively.

With regard to livestock, one cross bred cow is recommended in models M_2, M_4, M_5, M_6 and M_7 models and twelve sheep for models M_1, M_4 and M_7 models and four pigs in the case of M_5 model.

Small farmers

With respect to the objective of maximization of net income, the farmers could increase their net income from Rs.9,000 to Rs.11,650 (29.44%), Rs.17,125 to Rs.20,536 (19.92%), Rs.20,667 to Rs.24,280 (17.48%), Rs.42,800 to Rs.49,788 (16.33%) and Rs.49,250 to Rs.54,220 (10.09%) from models S_1, S_2, S_3, S_4, and S_5 respectively, in the normative plan over the existing ones (Table 4 & fig. 2).

Figure 2
Comparison between Existing and Optimal net Income under Maximisation for small Farmers

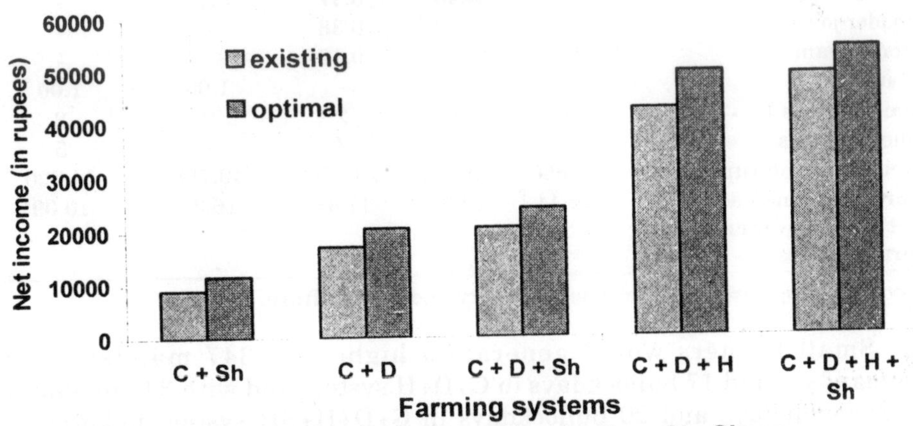

Note: C=Crop, D=Dairy, H=Dryland Horticulture Sh=Sheep

According to the normative farm plans, farmers are required to cultivate 0.36 ha and 0.22 ha of HYVs of groundnut + red gram in S_1 and S_5 models, and 0.44 ha and 0.56 ha of HYVs of groundnut + red gram + field bean in S_2 and S_3 models respectively, whereas HYVs of ragi + red gram with 8:2 ratio was suggested to be taken up in 0.89 ha, 0.56 ha, 0.26 ha and 0.33 ha in S_1, S_2, S_4 and S_5 models. Fodder maize was suggested in 0.48 ha 0.37 ha and 0.11 ha for models S_2, S_3 and S_4, and fodder jowar in 0.38 ha for S_3 model, horse gram in 0.22 ha and 0.16 ha for S_1 and S_3 models and mango in area of 1.00 ha each in S_4 and S_5 models.

With regard to livestock, two cross bred cows are suggested in each of the S_2, S_3, S_4, and S_5 models, seven sheep in S_1 and S_3 models and five sheep in model S_5.

The quantum of employment opportunities from various farming systems among marginal and small farmers in the normative farm plan is presented in the table 5. It could be observed that in the case of marginal farmers, C+D+H+Sh system provided highest employment to the extent of 360 mandays and 220 womandays followed by C+Sh+D with 292 mandays and 180 womandays of employment. In addition to family labour, each farmer has to hire both human and bullock labour in performing various operations in the normative farm plan.

Table 5
Normative farm plans for Maximization of net Income of small farmers

(area in hectares)

Farming System Activity	*C+ShS₁*	*C+DS₂*	*C+D+ShS₃*	*C+D+HS₄*	*C+D+H+ShS₅*
Groundnut + red gram (8:2)	0.36	-	-	-	0.22
Groundnut + red gram + field bean (8:2:1)	-	0.44	0.56	-	-
Ragi + red gram (8:2)	0.89	0.56	-	0.33	0.26
Fodder maize	-	0.48	0.37	0.11	-
Fodder jowar	-	-	0.38	-	-
Horse gram	0.22	-	0.16	-	-
Mango	-	-	-	1.00	1.00
Dairy (in nos.)	-	2	2	2	2
Sheep (in nos.)	7	-	7	-	5
Net income (in rupees)	11,650	20,536	24,280	49,788	54,220
Percentage increase in net income over existing farming system	29.44	19.92	17.48	16.33	10.09

Note: C=Crop, D=Dairy, Sh=Sheep, H=Dryland Horticulture.

Small farmers would generate a highest of 347 mandays, 298 womandays and 17 bullockdays in C+D+H system and with 340 mandays, 312 womandays and 20 bullockdays in C+D+H+SH system followed by C+SH+D system.

It is interesting to note that both the marginal and small farmers would gain additional employment opportunities in some of the normative plans developed. Marginal farmers would be benefited with additional employment of human and bullock labour in C+D+H (80 mandays and 57 womandays) and C+D+H+SH (60 mandays) systems while small farmers would gain the same in C+SH (80 womandays and 3 mandays) and C+D (81 womandays) systems.

Table 6
Labour Employment in the Normative farm plans among Marginal and small farmers

(in days)

Category *Farming systems*	*Marginal farmers*			*Small farmers*		
	M	*W*	*BP*	*M*	*W*	*BP*
C+Sh	220	180	11	212 (3)	200 (80)	9
C+D	119	161 (1)	-	184	212 (81)	15 (1)
C+H	65	50	-	-	-	-
C+Sh+D	300	200 (10)	-	309	175	10
C+D+Pi	168	144	-	-	-	-
C+D+H	292 (80)	180 (57)	-	347	298	17
C+D+H+Sh	360 (60)	220	-	340	312	20

Note: 1. C=Crop, D=Dairy, Sh=Sheep, H=Dryland Horticulture, Pi=Piggery.
2. M=Man, W=Woman, BP=Bullock pair.
3. Figures in the parentheses indicates increase in employment days over the existing farming systems.

CONCLUSIONS

It can be concluded that the introduction of developed farming systems through normative (optimum) farm plans appear to produce only modest income gains (less than 20%) for small farmers and the income increased found to be more than 23 per cent for marginal farmers. There was high degree of resource use inefficiency in dryland farms that can be ascribed to seasonality of agricultural crops i.e., rainfed agriculture is possible only during kharif season and in other seasons resources were mostly inoperative. The existing farming systems regardless of the farm size are less efficient ones as the farming systems developed through normative farm plans exhibited potential for higher income and employment.

REFERENCES

FAO Year Book (2000), Food and Agricultural Organization of the United Nations, Rome.

Kanwar, J.S. (1999), Need for a Future Outlook and Mandate for Dryland Agriculture in India in Fifty Years of Dryland Agricultural Research in India, *Central Research Institution for Dryland Agriculture*, Hyderabad, pp. 11-19.

Nagaraja, G.N. (1995), "A Study of Farm Business Management in Rural District of Karnataka: Compromise Programming Approach" Ph.D. *Thesis (Unpublished),* Tamil Nadu Agricultural University, Coimbatore.

NBSSLUP (2001), Classification of Dryland Region in India, *National Bureau of Soil Survey and Land Use Planning*, Nagpur.

Rajeswari, K. (2003), Sustainable Farming System Models for Drylands: An Economic Study in Kolar District of Karnataka, M.Sc., *Thesis (Unpublished),* University of Agricultural Sciences, Bangalore.

35

FARMERS' PARTICIPATION IN IRRIGATION MANAGEMENT —FUNCTIONING OF WUAS FORMED BY CADA IN HARYANA

C.S. Singhal

It has been widely accepted now that participation of farmers is imperative in the management of irrigation. National water Policy 1987 stresses on the progressive participation of farmers in the management of irrigation. Under centrally sponsored Command Area Development (CAD) program, participating states since 1986 are suppose to encourage participation of farmers in the management of irrigation water. As an incentive, management subsidy is being provided for the area to be covered under participation of farmers. However very little progress has been made so far. Andhra Pradesh have earned the distinction of becoming the first state in the country to hand over the Operation and Maintenance of the system at minor canal level to WUAs in entire command.

IRRIGATION MANAGEMENT IN HARYANA

The state of Haryana lies between the two river basins of Indus and Ganges and has a geographical area of 43.91 lakh hectares with a cultivable area of 38.27 lakh hectares. The rainfall in the State is scanty and erratic varying annually from as low as 300 mm in the South Western parts, border of Rajasthan, to about 1100 mm at the foothills of Himalayas in North Eastern region. No perennial river passes through the state. The river Yamuna runs along the boundary between U.P. & Haryana. Water resources of the Haryana State are limited.

IRRIGATION UNDER CANAL SYSTEM

At present 92 per cent of the cultivable land (35.8 lakh hac.) is the cropped area of which around 77 per cent is irrigated (26.88 lakh hac.). Canal system provide irrigation to about 52 per cent of the area, tubewell covers 47 per cent and 1 per cent is served by other sources like tanks and ponds etc.

There are three major canal system in the state. The first one is that Bhakra canal system, Which is fed from the Bhakra Dam, built across river Satluj. The second is the Western Yamuna Canal system. Which receives supply from river Yamuna at *Tajewala*. The third system is Gurgaon and Agra canal system, which again draws water from river Yamuna at Okhala. The south and south western parts of the state do not have the benefit of these canal systems. However, a series of lift Irrigation process are constructed in this area the supplies for these projects are through WYC system. Water resources agencies.

The Haryana Irrigation Department (HID) has the primary responsibility for managing the state's water resources and provides essential services in this sector. Two other agencies of Govt. of Haryana are also involved in irrigation sector. The first one is the Haryana State Minor. Irrigation and Tubewell Corporation (HSMITC), established in the 1970. The HSMIT was assigned to takeup activities related to ground water development under public sector through deep tube well. It also does the lining of water courses in canal irrigated command areas. The HSMTC has been closed down in the year 2002. The Command Area Development Agency (CADA) is the second organization, established in 1975, it does lining of watercourses where irrigation utilization is lagging, particularly, under lift irrigation schemes.

MANAGEMENT OF CANAL IRRIGATION WATER

The HID has the responsibility for the O & M of canals, distributories minors and outlets up to head of watercourses which are the lowest common channels serving an area of 100-300 ha. According to the existing Haryana Canal Drainage Act 1974, the maintenance and repair of water courses is the responsibility of the farmers; alternatively, the Act provides for the department to operate and maintain the system and charge *farmers* sufficiently to cover the cost. This however is not happening due to various reasons.

The farmers are maintaining the *Katcha* watercourse either individually in the length passing through their fields or through informal group actions, where some leading farmers take the initiative to organize other farmers to clear the channels. In case of lined watercourse, even though the farmers have to maintain them, over the years the responsibility is gradually taken over by the government.

Operation of irrigation system for water supplies are done in two phases, one of that is Govt. managed and the second is farmer managed. In the first phase, available water for irrigation is distributed in accordance with pre-defined priorities, major canals are grouped in to three areas and each area is assigned a "preference order" which defines the rights of that group in relation to the other two groups.

The distribution of water coming out of an outlet and flowing into a watercourse is the second phase of distribution and is managed by cultivators as provided under Secction 55 of the Haryana Canal and Drainage Act, 1974. The distribution of the water is done on a seven-day rotation basis, with the help of an approved roster, which divides 168 hrs. in the ratio of the holdings.

The rotation for drawal of water by each shareholder is fixed well in advance, which is called *"Warabandi"* and is informed to the farmers. *"Warabandi system"* of distributing irrigation water, from a watercourse has its main thrust as equity and one of the essential feature of this system is that all farmers, irrespective of the location of their holding in the command, are entitled to receive water in proportion to the size of their holdings which is reflected in terms of flow time allowed to each shareholder. Thus, land and not the cropping pattern is the base for the allocation of water.

Any dispute regarding preparation and operation of Warabandi or roster is settled generally by the farmers themselves. The divisional canal officer, who has quasi-judicial power to decide the dispute and to enforce the decisions, will come into the picture, when farmers are not able to resolve the dispute by themselves.

In the existing system of management of irrigation water, it is directly controlled and managed by the state, itself. The beneficiaries have virtually neither any collective sense of ownership to the system, nor any feeling that they should look after it. Hence Govt. of Haryana has started involving farmers by formation of Water User Association under PIM.

PARTICIPATORY IRRIGATION MANAGEMENT IN HARYANA

As a part of the implementation of the water Resources Consolidation Project (WRCP) started in 1994, the Government of Haryana introduced PIM from February 1995. A gradual turnover of the irrigation system at the water course level to the Water User Associations for operation and maintenance is the policy of the state Government. The Participatory Irrigation Management attempts to organise farmers and mobilise their skills and resources to encourage them to take an active role in operation and maintenance of watercourses. It has been decided to give priority to rehabilitation and modernisation of watercourses where farmers are willing to form WUAs and take over maintenance responsibilities. For effective functioning of WUAs, the state Government has approved the draft Bye-Laws and MoU in October, 1998.

Out of 6000 watercourses a total of 2079 WUAs have been registered under the Societies Registration Act, 1860. 991 WUAs registered were formed by under the guidance of Command Area Development Agency out of which 227 handed over to farmers. 1088 WUAs registered out of which 149 handed over to farmers were formed under the guidance of Haryana State Minor Irrigation and Tube wells Corporation and Haryana Irrigation Department.

OBJECTIVES

The main objectives of the present study was to know the perception of farmers & officials about the functioning of WUAs; study the farmer's participation in WUAs; and identify the constraints & suggest measures for effective implementation of Participatory Irrigation Management.

Study Area confined to 13 WUAs formed by CADA. These WUAs were selected in consultation with concerned officials of CADA.

Table 1 shows the list of WUAs under study command wise. Ten WUAs were studied under Western Jamuna Canal Command, three were under Agra Canal Command.

Table 1
List of WUA under Study Command wise

Name of command	*No.*	*Name of WUA*
CADA		
1. Western Jamuna Canal	10	1. Bhagwatipur, 2. Bhalli Anadpur, 3. Bhavar, 4. Bhavar-Nandgarh, 5. Gharwal, 6. Jeetgarh Roopgarh 7. Kathura, 8. Makrauli, 9. Quila Zafar garh, 10. Sunderpur.
2. Agra-Gurgaon Canal	3	1. Agvanpur-Firozpur, 2. Dighat Raidaska, 3. Patti Malian..

Table 2
Profile of WUAs Under Study formed by CADA

Sr. No.	*Name of WUA District wise*	*Name of Channel*	*No of Villages*	*RD of outlet*	*discharge (cs)*	*CCA (ac)*
1.	FARIDABA					
	a. Agvanpur-Ferozpur	Palwal Distb.	2	17-R	.72	231
	b. Dighot -Raidaska	Hodal Distb.	2	66-R	.92	296
	c. Patti-Malian	Palwal Distb.	1	28-R	.52	135
2.	JIND					
	a. Quila Zafar Garh	QZH Minor	1	49025 -T	1.67	698
	b. Bhawar Nandgarh	QZH Minor	2	3078 -R	.77	347
	c. Bhawar	SunderSubBr.	1	30384 -L	2.83	853
	d. Jeetgarh-Roopgarh	Hansi Branch	2	82470 -R	.60	249
3.	PANIPAT					
	a. Kathura	Kathura minor	1	15120 -R	1.34	559
	b. Gharwal	Butana minor	1	113955-L	1.72	1000
4.	ROHTAK					
	a. Makrauli	Makrauli minor	1	13500	1.73	722
	b. Sunderpur	Rohtak Distb	1	120150-R	1.70	662
	c. Bhagratipur	Kahnaur Distb	1	67050 -L	1.62	676
	d. Bhali Anandpur	Kahanaur Distb.	1	109800-L	0.76	318
Total	6746 ac					

R.D. outlet (Reduced Distance) - Distance of the outlet from the head of minor / distributory; L = left to the flow of water; R = Right to the flow of water; TC = Tail Centre

PROFILE OF WUAS UNDER STUDY

All together 13 Water User Associations formed by CADA were studied. The 13 WUAs studied covered 16 villages spread in four district i.e.

Faridabad, Jind, Panipat and Rothak. Table 2 Shows the profile of WUAs of CADA under study. It shows that 13 WUAs covered CCA of 6746 ac. The rate of discharge varies from 0.52 cs in patti-Malian to 2.83 cs in Bhawar. Table 2 shows the profile of the WUAs under the study.

METHOD OF DATA COLLECTION & ANALYSIS

The data were collected through Participatory Rural Appraisal method from members of WUAs, executives of the Association, few influential people of the village, members of Panchayat. The small, marginal farmers, schedule caste & backward caste were also covered in the sample. About 15-20 people from each WUA area were contacted & interviewed individually & in-group after establishment of good rapport with them.

EVOLUTION OF WUAs

The Govt. of Haryana decided to give priority to rehabilitation and modernization of watercourses where farmers are willing to form WUA and take over maintenance responsibility. As a result of this the officials of CADA used to take over initiative information of WUA. They visited villages, identified interested farmers and with the help of them organized meetings of the farmers. During these meetings they explained about the advantages of constructed field channels and their role in operation and maintenance of these channels. It was also informed that there is prerequisite condition to form WUA and deposit share money @ Rs. 20/ac for construction of pucca watercourse.

Table 3
Name, Address of WUAs taken in the Present study

Name of WUAs	*Abbreation*	*Address*
CADA		
1. Agvan Pur-Feozpur	AFR	Villgaes-Agvanpur, Firozpur & Palwal Distt. Faridabad
2. Bhali-Anandpur	BAR	Bhali Anandpur, Dist. Rohtak
3. Bhawar	BHR	Bhawar, Dist. Jind
4. BhawarNanadgarh	BNH	Village Bhawar and Distt. Jind
5. Bhagwatipur	BPR	Village Bahwatipur, Distt.Rohtak
6. Dighot –Raidaska	BNH	Viilgae Dighot & Riadask, Town Palwal, Distt. Faridabad,
7. Gharwal	GHL	Village Ghawrwal, Dist.-Sonipat
8. Jeetgarh–Roopgarh	JRH	Villages Jeetgarh & Ropgarh Distt. Sonipat,
9. Kathura	KRA	Village Kathura, Gahana Block, Distt. Sonipat.
10. Makrauli	MKI	Village Makruali Distt. Rohatak
11. Pattimalian	PMN	PattiMalian, Town Palwal, Distt. Fardabad,
12. Qulia -Zafargarh	QZH	Village Quila Zafargerh, Dist. Jind,
13. Sunderpur	SUR	Village Sunderpur Distt. Rohtak

With the continuous effects of CADA Officials the farmers were convinced ultimately. Initially, although they were reluctant and having apprehension about the proper use of their share money. When officials explained to them that the share money of farmers and the matching grant from the govt. would be placed in fixed deposit account and the interest will be used by the members of WUA for repairing the watercourses. The farmers were convinced. In the presence of officials they formed the association and selected unanimously President, Secretary, Cashier and few Executive Members as Managing Committee.

The motivation level was quite high initially in most of the Associations because the farmers realised that pucca field channels will be saving wastage of water and it will be easy to operate. They were of the view that theft will also be reduced considerably and due to increased flow, the tail enders will also get sufficient irrigation water. As a result of this realisation the farmers has deposited their share money.

The Managing Committee members hold informal meetings as and when required. However, they did not record the proceedings regularly. The decisions were taken democratically and unanimously.

SHARE HOLDERS PARTICIPATION IN PLANNING STAGE

The shareholders were fully involved in selection of site. There was no dispute regarding this matter in most of the Associations, expect in village Agvanpur-Firozpur where there was some dispute between farmers of two villages, which was ultimately settled.

The construction of channel was executed by private contractor under the guidance of CADA officials.

The farmers were also involved in monitoring the quality of construction material and lining work in most of the villages expect in village Kathura, Agvanpur- Firozpur, Sunderpur and Bhagvatipur, Where the participation of farmer was not up to the mark.

In villages Bhawar, Gharwal and Jeetgarh-Roopgarh the shareholders even pressuried contractor to replace the sub-standard quality of bricks. Normally the Managing Committee deputes three members to keep watch on the lining work. They remain on the site turn by turn. After Construction the watercourses were handed over to farmers.

THE SHAREHOLDERS PARTICIPATION IN OPERATION & MAINTENANCE

The water was distributed as per warabandi system in most of the cases expect in Agvanpur-Firozpur where it was not fixed. Because of non-existence of fixed time there was conflict among farmers. Hence the farmers suggested to fix the time. It was observed that farmers co-operate normally in distribution of water. There was no dispute in this regard in most of the cases.

The shareholders were cleaning the silt and vegetation from the watercourse just before the discharge of water in most of the Associations.

The society has made a fine provision of Rs. 50 in the village Bhawar. However it was not implemented rigorously.

In Bhagwatipur and Sunderpur farmers were not caring the watercourse rather they were doing harm to the structure for getting more water. In Sunderpur since there was no water available in the distributory, the farmers were taking water from tubewell and drain through watercourse and thereby damaging it.

The farmers of Bhagwatipur realize the importance of watercourse and during study visit they wanted to get it repaired out of share money.

IMPACT

The water use efficiency has been increased due to pucca channel in most of the cases. The farmer revealed that the area under irrigation has increased from 25 to 50 per cent. The tail enders started getting increased irrigation water.

As a result of increased water availability the cropping pattern has started changing in most of the cases. Earlier the paddy was grown only near by channel but now the farmers started growing paddy in other areas also.

It was revealed by farmers that after pucca watercourses the wheat production has increased by 2/3 Quintals per acre. The sugarcane production increased about 4 to 5 quintal per acre.

The farmers were aware about their roles & responsibilities but they were not very clear about the Bye Laws & MoU.

It was found that there was no woman participation in the activity of WUA. Except in Gharwal and Kathura WUA were there was only one women member in the Managing Committee. In rest of the cases, there was no women representation. These two women also never participated in any of the meetings. The main reason of their non-participation was social stigma, purdah system and illiteracy.

The WUA is independent registered body has no linkages with the village Panchayat. The villagers were neutral regarding the involvement of Panchayat. However, they were looking Panchayat as more powerful in comparison to WUA. They wanted more powers and recognition to the Association like village Panchayat.

CONSTRAINT AND SUGGESTIONS

1. The most of the Shareholders were not aware about the details of By-Laws and MoU. They even not organize General Body and Managing Committee meetings as prescribed in the Bye Laws. They do not follow the procedure of maintaining proper records. Hence the shareholders should be trained periodically by the officials of irrigation department and CADA.
2. The most of the middle level officials of CADA and irrigation department were not very clear about their responsibility to strengthen the functioning of WUA. They were mainly concerned upto the

formation of WUA and collection of share money so that there target of lining could be achieved. After formation of WUA there was hardly any meeting with the share holders. Hence all the officials of CADA, irrigation department and other concerned department should be trained periodically to make them clear about there responsibilities to strengthen the effective functioning of WUAs.

3. In most of the cases the farmers revealed that they are not getting any guidance and assistance regarding water use efficiency from any of the development department like agriculture, horticulture, soil conservation, forest. Even in two-three cases where agriculture development officers visited the villages, the benefit was not reaching to small and marginal farmers. Hence, simultaneously at the time of lining of watercourses extensive services should be provided for optimum utilization of water and increase in production.
4. It was observed that large numbers of areas in most of the cases remains to be covered under lining. The shareholders whose field are not covered were demotivated. Hence the present norm of 24'per acre lining needs to be revised to at least 35 to 40 feet per acre as suggested by farmers.
5. The shareholders were critical about the low discharge of water and were quite concerned with this issue. Because of low discharge the warabandi time was not sufficient. Hence, they requested to enhance the discharge rate and Warabandi time.
6. It was revealed by shareholders of villages Bhagwatipur, Bhali Anandpur and Bhavar- Nandgarh that in some places *pannels (Nakhas)* were not fixed. They requested that there *Nakhas* should be fixed on priority to avoid in- convenience in running the water.
7. In the most of the cases there was problem of leakage and over flow in watercourses. The reason of over-flow was improper levelling of watercourse bed, less worth of wall and lack of plastering on the wall. Hence it is suggested that to control leakage atleast one layer of plaster inside the watercourse should be made. Wherever there is overflow the height of the wall may be raised.
8. The shareholders of Sundarpur are loosing their interest in the operation and maintenance of watercourse due to non-availability of canal water for the last 2 years. Hence efforts should be made to get restored the water in the canal.
9. The interest earned on the funds of WUA will be helpful in minor repair only. For major repair works the Association will require additional funds. Hence, Associations should be guided to mobilize more funds. Moreover there should be flexibility in utilization of the existing funds. There should be clear guidelines formed in this regard.

REFERENCES

Brewer, J.D. & others (1997), *Irrigation Management Transfer in India: Policies, Processes & Performance.* IIM Ahemdabad, IIM Colombo.

Gupta, Deepak & Srivastava, L.P. (1999), *Status of Participatory Irrigation Management in India,* New Delhi: Institute of Social Sciences.

IRDAS (1997), Water User Associations, *PIM–Haryana, Orientation Training Programme, HIRMI, Kurukshetra,* IRDAS, Hyderabad.

IRDAS (1997), *Evolving strategy for Implementation of Pilot Programme Vol. I Pilot Prog. for Establishing WUAs-Haryana,* GOH Dept. Irrigation, Chandigarh.

——— (2004), *Participatory Irrigation Management–A Case Study of WUAs in Haryana,* Research Report, NIRD, Hyderabad.

——— (2002), *Participatory Irrigation Management: A Case Study of Haryana Village,* "Emancipation of Masses for Social Development", S. Vijya Kumar, D. Vasudeva Rao, R. Venkata Ravi (ed.), Serial Publications, New Delhi.

———(2001), "Emerging Institutions of Irrigation management: A Case Study of Quila Zafargardh WUA in Haryana", *Prestige Journal of Management and Research,* V (1).

———(2001) *"Participatory Irrigation management: A Study of WUAs in Haryana"* "Private Extension: Indian Experiences, National Institute of Agriculture Extension Management", Hyderabad, 81-89.

———(2000), *Participatory Irrigation Management–A Study of WUAs in Haryana,* Nilokheri, Karnal, HIRD, Research Report, p. 250.

36

PARTICIPATORY IRRIGATION MANAGEMENT—A CASE STUDY OF BHAWAR WUA IN HARYANA

C.S. Singhal

National water Policy 1987 stresses on the progressive participation of farmers in the management of irrigation. The participation of farmers is imperative in the management of irrigation. Under centrally sponsored Command Area Development (CAD) program, participating states since 1986 are suppose to encourage participation of farmers in the management of irrigation water. As an incentive, management subsidy is being provided for the area to be covered under participation of farmers. However very little progress has been made so far. Andhra Pradesh have earned the distinction of becoming the first state in the country to hand over the Operation and Maintenance of the system at minor canal level to WUAs in entire command.

IRRIGATION MANAGEMENT IN HARYANA

The state of Haryana lies between the two river basins of Indus and Ganges and has a geographical area of 43.91 lakh hectares with a cultivable area of 38.27 lakh hectares. The rainfall in the State is scanty and erratic varying annually from as low as 300 mm in the South Western parts, border of Rajasthan, to about 1100 mm at the foothills of Himalayas in North Eastern region. No perennial river passes through the state. The river Yamuna runs along the boundary between U.P. & Haryana. Water resources of the Haryana State are limited.

IRRIGATION UNDER CANAL SYSTEM

At present 92 per cent of the cultivable land (35.8 lakh hac.) is the cropped area of which around 77 per cent is irrigated (26.88 lakh hac). Canal system provide irrigation to about 52 per cent of the area, tubewell covers 47 per cent and 1 per cent is served by other sources like tanks and ponds etc.

There are three major canal system in the state. The first one is that Bhakra canal system, Which is fed from the Bhakra Dam, built across river Satluj. The second is the Western Yamuna Canal system. Which receives supply from river Yamuna at *Tajewala*. The third system is Gurgaon and Agra canal system, which again draws water from river Yamuna at Okhala. The south and south western parts of the state do not have the benefit of these canal systems. However, a series of lift Irrigation process are constructed in this area the supplies for these projects are through WYC system. Water resources agencies.

The Haryana Irrigation Department (HID) has the primary responsibility for managing the state's water resources and provides essential services in this sector. Two other agencies of Govt. of Haryana are also involved in irrigation sector. The first one is the Haryana State Minor. Irrigation and Tubewell Corporation (HSMITC), established in the 1970. The HSMIT was assigned to takeup activities related to ground water development under public sector through deep tube well. It also does the lining of water courses in canal irrigated command areas. The HSMTC has been closed down in the year 2002. The Command Area Development Agency (CADA) is the second organization, established in 1975, it does lining of watercourses where irrigation utilization is lagging, particularly, under lift irrigation schemes.

MANAGEMENT OF CANAL IRRIGATION WATER

The HID has the responsibility for the O & M of canals, distributories minors and outlets up to head of watercourses which are the lowest common channels serving an area of 100-300 ha. According to the existing Haryana Canal Drainage Act 1974, the maintenance and repair of water courses is the responsibility of the farmers; alternatively, the Act provides for the department to operate and maintain the system and charge *farmers* sufficiently to cover the cost. This however is not happening due to various reasons.

The farmers are maintaining the *Katcha* watercourse either individually in the length passing through their fields or through informal group actions, where some leading farmers take the initiative to organize other farmers to clear the channels. In case of lined watercourse, even though the farmers have to maintain them, over the years the responsibility is gradually taken over by the government. Operation of irrigation system for water supplies are done in two phases, one of that is Govt. managed and the second is farmer managed. In the first phase, available water for irrigation is distributed in accordance with pre-defined priorities, major canals are grouped in to three areas and each area is assigned a "preference order" which defines the rights of that group in relation to the other two groups.

The distribution of water coming out of an outlet and flowing into a watercourse is the second phase of distribution and is managed by cultivators as provided under Secction 55 of the Haryana Canal and Drainage Act, 1974. The distribution of the water is done on a seven-day rotation basis, with the help of an approved roster, which divides 168 hrs. in the ratio of the holdings.

The rotation for drawal of water by each shareholder is fixed well in advance, which is called "*Warabandi*" and is informed to the farmers. "*Warabandi system*" of distributing irrigation water, from a watercourse has its main thrust as equity and one of the essential feature of this system is that all farmers, irrespective of the location of their holding in the command, are entitled to receive water in proportion to the size of their holdings which is reflected in terms of flow time allowed to each shareholder. Thus, land and not the cropping pattern is the base for the allocation of water.

Any dispute regarding preparation and operation of Warabandi or roster is settled generally by the farmers themselves. The divisional canal officer, who has quasi-judicial power to decide the dispute and to enforce the decisions, will come into the picture, when farmers are not able to resolve the dispute by themselves.

In the existing system of management of irrigation water, it is directly controlled and managed by the state, itself. The beneficiaries have virtually neither any collective sense of ownership to the system, nor any feeling that they should look after it. Hence Govt. of Haryana has started involving farmers by formation of Water User Association under PIM.

PARTICIPATORY IRRIGATION MANAGEMENT IN HARYANA

As a part of the implementation of the water Resources Consolidation Project (WRCP) started in 1994, the Government of Haryana introduced PIM from February 1995. A gradual turnover of the irrigation system at the water course level to the Water User Associations for operation and maintenance is the policy of the state Government. The Participatory Irrigation Management attempts to organise farmers and mobilise their skills and resources to encourage them to take an active role in operation and maintenance of watercourses. It has been decided to give priority to rehabilitation and modernisation of watercourses where farmers are willing to form WUAs and take over maintenance responsibilities. For effective functioning of WUAs, the state Government has approved the draft Bye-Laws and MoU in October, 1998.

Out of 6000 watercourses a total of 2079 WUAs have been registered under the Societies Registration Act, 1860. 991 WUAs registered were formed by under the guidance of Command Area Development Agency out of which 227 handed over to farmers. 1088 WUAs registered out of which 149 handed over to farmers were formed under the guidance of Haryana State Minor Irrigation and Tube wells Corporation and Haryana Irrigation Department.

OBJECTIVES

The main objectives of the present study was to know the perception of farmers & officials about the functioning of WUA ; study the farmer's participation in WUA ; and identify the constraint & suggest measures for effective implementation of Participatory Irrigation Management.

The *Bhawar* Water User Association was formed under the guidance of CADA and registered in 1997 under the Societies Registration Act, 1860. It covered one village is situated in Jind district of Haryana State (India). The total shareholders were about 254, out of which big farmers were 104, followed by small (100) and marginal farmers (50). Except 10-12 members who were OBC, all others belonged to general caste i.e. *Jaat* Community. Managing Committee comprised of 7 farmers. Out of which six were small and one marginal farmer. The Association was formed under the guidance of CADA Lg. Division, Jind. The WUA was responsible for operation & maintenance of watercourse below outlet level. The RD of outlet was 30384-L with rate of discharge 2.83 cs. The CCA of outlet was 853 ac. The water was supplied to watercourses through Sunder Sub Branch of Bhakra Canal. The details of WUA are given in Table 1.

Table 1
Profile of Bhawar WUA

Sr. No.	*Item(s)*	*Detail*
1.	Villages Covered	One Bhanwar
2.	Year of Registration	1997-98
3.	No. of Shareholders	254
	(a) Big Farmers (> 5ac.)	104
	(b) Small Farmers (2.5 to 5 ac.)	100
	(c) Marginal Farmers (< 2.5 ac.)	50
4.	Managing Committee Members	7
	(a) Big Farmers (>5ac.)	6
	(b) Small Farmers (2.5 to 5 ac.)	1
	(c) Command Farmers (<2.5 ac.)	-
5.	Command	Western Jamuna Canal
6.	Channel	Sunder Sub Branch
7.	RD of outlet	30384 - L
8.	Rate of Discharge	2.83 cs
9.	C.C.A. of Outlet	853 ac.

PROFILE OF VILLAGE

The *Bhawar* village is about 25 Km. from district Jind. Its geographical area was about 2000 ac. and command area 1640 ac. it has about 1980 households with population of 10,000. The majority of people belonged to Jaat Comunity (1150).There were Brahmin (230) with considerable number of SC (350) and OBC (250).

OCCUPATION

The main occupation of villagers was Agriculture. Small farmers (700) were in large number, followed by marginal (625) and big farmers (50). About 600 families were landless, engaged in daily wages and services.

IRRIGATION SOURCES

The main sources of irrigation were Sunder-Sub-branch and Quila Zafargarh Minor, originated from Western Jamuna Canal and private tubewell (200). Tubewells were not used in excess due to salinity in ground water. Table 2.

Table 2
Profile of Village Bhawar (District Jind)

Sr. No.	*Item(s)*	*Details*
1.	Total Households	1980
	(i) General	1380
	(ii) OBC	250
	(iii) SC	350
2.	Total Population	10,000
3.	Literacy Rate (%)	45%
4.	Village Geographical Area (ac.)	2000
5.	Total Command Area (ac.)	1650
6.	Land Holding	
	(a) Big Farmers (> 5 ac.)	50
	(b) Small Farmers (2.5 to 5 ac.)	700
	(c) Marginal Farmers (<2.5 ac.)	625
	(d) Landless	600
7.	Source of Irrigation	
	(a) Tubewell	(i) Saline Water (50) (ii) Mixed (150)
	(b) Ponds	(i) With Water -2
	(c) Channels	Sunder Sub Branch
8.	Common Land (ac.)	14
9.	Live Stock Population	(i) Buffalo - 4000 (ii) Cow - 200 (iii) Goat/Sheep - 200 (iv) Bull - 100
10.	Local Institutions	(i) Govt. School - 2

EVOLUTION OF WUA

The Association came into existence as a consequence of joint efforts of farmers and officials of CADA. They organised several meetings with villagers and explained to them about the advantage and objective of forming Water User Association as operation and maintenance of watercourse and distribution of water equitably and more efficiently. Farmer's contribution at the rate of Rs. 20/ac. was also clarified to them with matching grant to be kept in fixed deposit in Association's name.

Almost all the shareholders attended the meetings organised by officials of CADA. The motivation level was high among farmers. Managing Committee was also formed with the consent of farmers. They further cooperated with the CADA officials in organising meeting for further motivating the farmers. Few committee members were educated and aware about the Bye-Laws and their responsibility. Farmers also actively participated gave, their consent and views in democratic way.

Regarding the matter of contribution almost all of farmers gave their share money. The wholesome amount including the contribution of farmers as well as government i.e. Rs. 1,52,500 had been deposited in the bank in fixed deposit. Three members of Managing Committee, President, Secretary a Cashier operate the bank account. The cashier possesses Rs. 1,000/-, which could be used by association. Records were also kept by the committee members. After the collection of share money the lining of *pucca (cemented)* watercourse has started. The construction work was got executed by CADA, through contractor. During the construction work farmers fully participated. They checked the material and progress of construction. The Association deputed a watchman for continuous vigilance and also paid remuneration of Rs. 1500/- per month. Other farmers also kept their eyes on the construction work. Once they returned the construction material back which was not of good quality, and also replaced the same. Farmers were critical of substandard quality of construction material and non-implementation of standard design as one of the reason of leakage of water from watercourse wall. Teamwork was satisfactory.

After the watercourses were handed over to them, farmers were very happy and liked the newly constructed watercourses. Farmers started getting water as per existing warabandi system i.e. 15 minutes for one acre.

IMPACT

The per unit availability of water was increased. The production was also increased due to increase in water use efficiency. After the formation of pucca watercourse the production of paddy and wheat was increased 8-10 quintal/acre of land. Although there was no change in cropping pattern. The main crop was wheat, sugarcane, paddy, cotton and millet.

MAINTENANCE OF WATERCOURSE

Farmers also participated in the maintenance of watercourse. The farmers used to go in their fields to clean the channel. They removed sand, silt and undergrowth from the watercourse. It was the duty of each farmer to keep the watercourse in neat and clean condition across his land. If any farmer did not fulfill his duty, he could be fined with Rs. 50/- as per rule made in WUA But it had not happened till now. They co-operated with each with other. If any farmer was not able to fulfil his task due to any reason, other farmers did his job.

Farmers told that water theft did not take place, as it was rather difficult task to steal water from lined watercourse.

Disputes regarding the distribution of water were denied by farmers. Committee members told that if it takes place, it would be resolved among themselves amicably. Role of Panchayat was neutral.

CONSTRAINT AND SUGGESTIONS

1. Farmers were facing the problem, reflected in the watercourse, like in some places the path of watercourse was not properly leveled. As a

consequence water flow got disturbed and over-flow occurred. The water came out of watercourse. It was suggested that additional layers of brick should be placed so that the height of walls of watercourse may be increased and over flow comes under control;

2. Though large area was covered, yet some land was still deprived of pucca field-channels. Hence it was suggested that remaining area should also be covered by further extension of pucca watercourse.
3. As the region comes under saline water zone, farmers wanted to get installed a Govt. tubewell nearby canal for collective use.
4. It was revealed that time allotted, as per warabandi system was not sufficient. Hence they suggested increasing the time. They wanted discharge more convenient.
5. The farmers wanted that they should be allowed to draw their amount timely for maintenance and operation of watercourses.

REFERENCES

Brewer,J.D. & others (1997), *Irrigation Management Transfer in India: Policies, Processes & performance.* IIM Ahemdabad, IIM Colombo.

Gupta, Deepak & Srivastava, L.P., (1999), *Status of Participatory Irrigation Management in India,* New Delhi: Institute of Social Sciences.

IRDAS (1997), Water User Associations, *PIM–Haryana, Orientation Training Programme, HIRMI, Kurukshetra,* IRDAS, Hyderabad.

IRDAS (1997), *Evolving strategy for Implementation of Pilot Programme Vol. I Pilot Prog. for Establishing WUAs-Haryana,* GOH Dept. Irrigation, Chandigarh.

Singhal,C.S. (2001), *"Participatory Irrigation management: A Study of WUAs in Haryana"* "Private Extension: Indian Experiences, National Institute of Agriculture Extension Management", Hyderabad, 81-89.

——(2000), *Participatory Irrigation Management–A Study of WUAs in Haryana,* Nilokheri, Karnal, HIRD, Research Report, p. 250.

37

SUSTAINABLE AGRICULTURE AS A PANACEA FOR RURAL POVERTY AND UNEMPLOYMENT

K.S. Venkateshappa

CONCEPT OF SUSTAINABILITY

Development economists have adopted the term sustainability or sustainable development which means development that meets the needs of the present generation without compromising the ability of future generations to meet their own needs[1]. But Bartelmus (1994) introduced a new definition focusing on the maintenance of produced and natural capital for ensuring continuing generation of output and value-added and bearing in mind possibilities of extending the use of natural capital through resource saving and environmentally sound technological progress, resource discovery of substitution of produced, natural or human capital inputs[2]. It is not a fixed state of harmony, but rather a process of change in which the exploitation of resources, the direction of investments the orientation of technological development, and institutional change are made consistent with future as well as present needs.

Rajkrishna's 'Hindu Rate of growth' of 4 per cent will not solve India's economic problems in the foreseeable future. Even twice of that rate of growth, which India has not historically achieved yet, will not solve the problem of appelling poverty and unemployment that exists in the country, for many, many years to came. To be able to achieve her rightful place in a highly competitive world within one generation of say, 25 to 30 years, India must seek to attain a long and sustainable (annual) 10 to 12 per cent rate of growth in her GDP by the end of *X* Plan.

During the past three decades, when the less developed countries of Asia, South America and Africa entered the phase of development, there was a strong bias in favour of rapid industrialisation. This was largely due to feeling among them that industrialisation was synonymous with economic development. However, the strategy that was adopted towards this end faced serious obstacles and several countries failed in their endeavors and ended in frustration. Historical records clearly show that no country has moved from the stage of chronic stagnation to the take-off stage of economic development without first achieving a substantial increase in agricultural production. Gunnar Hyrdal[3] concludes that it is in the agricultural sector that the battle for long-term economic development in South Asia will be won or lost. Even Coale and Hoover[4] with reference to India opines that very substantial progress in agriculture is a prerequisite to successful development of the Indian economy as a whole.

CONCEPT OF POVERTY

Poor people live without fundamental freedoms of action and choice that the better-off take for granted. They often lack adequate food and shelter, education and health, deprivations that keep them from leading the kind of life that everyone values. They also face extreme vulnerability to ill healthy, economic dislocation, and natural disasters. And they are often exposed to ill treatment by institutions of the state and society and are powerless to influence key decisions affecting their lives. These are all the dimensions of poverty. An alternative definition of poverty–expounded by the British Sociologist Peter Townsend, among others-is a lack of resources required to participate in activities and to enjoy living standards that are customary or widely accepted in the society in which poverty is being measured. Botswana Country study refers to poverty in terms of deprivation related to insufficiency in food, housing, health, education and other items required to maintain an acceptable standard or equality of life.

Poverty is the result of economic, political and social process that interact with each other and frequently reinforce each other in ways that exaccerbate the deprivation in which poor people live. Meager assets, inacessible markets, an scarce job opportunities lock people in material poverty. That is why promoting opportunity by stimulating economic growth, making markets, work better for poor people, and building up their assets–is keg to reducing poverty[5]. The existence of poverty in India cannot be denied. It is a matter of no surprise that millions of our countrymen remain inadequately fed, clothed and housed. Crores of people in the countryside live below the poverty line. The single most important reason for rural poverty is large scale unemployment in the rural sector. This attribute in itself explains the gigantic magnitude of the problem. The Indian study points out that the composition of growth appears to be important in explaining its impact on poverty reduction and that sustainable increase in agricultural output has a more pronounced effect on poverty reduction. Economic growth along with sustainable agricultural growth must 'trickle down' and it will also have a 'spread effect' in order to eradicated poverty and unemployment in India.

RURAL UNEMPLOYMENT

Prevalence of Poverty in rural areas is due to large scale unemployment and underemployment. Majority of ruralities are landless and sustains entirely on wage employment. They continue to suffer due to excessive seasonality of employment, lack of wage employment opportunities and low wage rates. Efforts to provide wage employment as short-run strategy for alleviation of rural poverty has been a major component of the development planning in our country.

Unemployment in the rural sector does not manifest itself in one single form. It often takes different forms and makes different manifestations. Unemployment in Indian rural sector is generally said to be of four major types. There is open unemployment, disguised unemployment, under employment, and seasonal unemployment. Massive investment made in the successive plans in other economic sectors has failed to take away surplus rural labour and the excessive pressure on land has resulted in the ever increasing sub-divisions and fragmentation of holdings. Further the growth of agricultural sector has also been slow with the result that this sector could not create enough opportunities for additional employment. The awareness of this failure has now generated a new national commitment to make a determined effort in the next decade to eliminate unemployment through sustained agricultural growth.

During the last 50 years, several schemes have been initiated to generate employment opportunities in the rural areas. All these programmes were launched exclusively for employment generation & poverty eradication in rural areas. But the dimensions of rural poverty in India are so variegated and its roots are so deep that a for more ambitious programmes are required to deal effectively with this malaise.

ECONOMIC GROWTH AND POVERTY

The evidence from the experience of a large number of countries is that while the impact of economic growth on reduction in inequalities is mixed. However, the benefits of growth are observed to be accruing to different economic groups approximately in proportion to their initial incomes. Further the composition of growth appears to be important in explaining its impact on poverty alleviation. It is generally agreed that growth propelled by an increase in agricultural output has a more pronounced effect on poverty alleviation.

In recent years, the positive relationship between growth in agricultural production and poverty alleviation seems to have weakened[6]. This is made clear from the Tables 1 and 2. Even then, the positive impact of agricultural growth on the income of the small farmers and more particularly on the wage income of the agricultural labourers cannot be denied. In the face of inequality in distribution of land holdings the positive impact of agricultural growth on poverty alleviation depends more on the capacity to generate employment. In this context growth in non-farm sector in the rural areas assumes importance. This is clearly brought out from the study of poverty alleviation in Haryana. Significant reduction in the rural poverty in Haryana. Employment opportunities in agriculture could be explained by a remarkable increase in the non-farm employment. The study of Andhra Pradesh has also brought out that poverty declined more rapidly in the districts adjoining Hyderabad City compared to those in the interior mainly because of the growth in the non-farm employment in the periphery of this city.

Table 1
Absolute Poverty: 1980-1990

Country	*Total*	*Urban*	*Rural*
Bangladesh	78	NA	86
India	40	33	42
Nepal	60	51	61
Pakistan	28	26	29
Srilanka	39	15	46

Source: The world Resource Institute (1996), World Resources 1996-97, New York, Data Table A-1.

Table 2
Food Production: 1980-92

Country	*Index of Food Production (1980-81-100)*			
	Total		*Per Capita*	
	1980-82	*1990-92*	*1980-82*	*1990-92*
Bangladesh	102	126	99	96
India	103	153	101	122
Nepal	102	156	100	116
Pakistan	104	159	101	112
Srilanka	102	99	100	84

Source: The World Resources Institute (1996), World Resources 1996-97, New York Data Table 2.

FUTURE FOOD PROBLEM

India has a buffer stock of about 35 million tonnes of foodgrains. The 1990s have seen a certain degree of diversification of agricultural and exports of various agricultural products including wheat and rice. There is also a growth in the agriculture–based processing industry. So can we rest on our own, comfortable in the belief that there are no more problems on the food front? But Yagnaswami Sundara Rajani's[9] Report on food and Agriculture; Technology vision 2020 the projected grain imports in 2000 and 2010 has been given in the Table 3.

According to table 3, India may have to important about 14 million tonnes of food grains by 2010 and the imports will grow at the rate of 2 per cent every year. Can we draw comfort from the fact that Pakistan will have to import 4.5 million tonnes in 2010 and China 21.6 million tonnes ? Since we need at least a 7 per cent growth rate to reach developed country status, it is safe to assume a demand of 349 million tonnes of food grains by 2020. All these projected increases in demand place additional pressures on Indian agriculture. The optimum allocation of land and other resources for various crops will itself pose a challenge. Can we declare, consume less milk or oil or eat less vegetables ? These are the new challenges before us in a not-too-distant future. A scenario for domestic demand for food grains for different rates of economic growth is given in Table 4. Along with many others who have studied these issues in depth and thought about possible solution, we believe that we need not accept these conclusions at all since India has tremendous potential for increasing production. Our people and our farmers are exceptionally entrepreneurial, and have proved it again and again.

Table 3
Projected grain imports in 2000 and 2010

Countries	*Million Tonnes*	
	2000	*2010*
South Asia	9.2	12.8
East Asia	31.4	39.0
India	6.9	14.1
Pakistan	2.1	4.5
Indonesia	5.7	7.6
China	11.3	21.6

Source: TIFAC, Food and Agriculture: Technology Vision 2020.

Table 4
Projected Household Demand for Food in India at 7 per cent Income Growth

Commodity	*Annual Household Demand (Million metric tones)*				
	1991	*1995*	*2000*	*2010*	*2020*
Foodgrains	168.3	185.1	208.6	266.4	343.0
Milk	48.8	62.0	83.8	153.1	271.0
Edible Oil	4.3	5.1	6.3	9.4	13.0
Vegetables	56.0	65.1	80.0	117.2	168.0
Fruits	12.5	16.1	22.2	42.9	81.0
Meat, Fish & Eggs	3.4	4.4	6.2	12.7	27.0
Sugar	9.6	10.9	12.8	17.3	22.0

Source: TIFAC, Food and Agriculture: Technology Vision 2020.

Thus the growing demand for food grains, vegetables, fruits, milk, poultry and meat as well as cash crops is going to present greater and newer challenges to agriculture. Let us not forget that our existing food security has been mainly brought about by the increase in irrigated agriculture and the introduction of high-yielding varieties of crops.

COMPETITIVENESS OF INDIAN AGRICULTURE

When India initiated economic reforms in 1991, domestic prices of several agricultural commodities were below international prices. This led to the conclusion that Indian agriculture is highly competitive and that freeing trade would help the country harness vast export potential. Global liberalisation of agriculture trade following the uruguay round of GATT was further seen as a great opportunity for promoting farm exports.

The situation has, however, changed dramatically after 1996. Domestic prices of most agricultural commodities have turned out to be higher than the international prices, and India has become attractive market for imports. What accounts for the fall in international agricultural prices during post- WTO period? Most of the developed countries, which were also major player in the international market, raised support to agricultural under the green fox category which exempt from reduction. Pressure is building up in WTO To treat all kinds of support to agriculture under one category and subject it to reduction as per the commitment. India should also seek and strongly support reduction in all kinds of subsidies and strike to provide level playing field to our agriculture.

BUDGET 2001-02

The Finance Minister's focus on agriculture has been for overdue. Speedy reforms in the sector have been a desire for a long, long while. The most heartening part of the budget is the vision of a national policy on agriculture. The budget promises a credit flow of 51,500 crores to agriculture this year and a further accentuated 64,000 crores in the next year, are numbers that most definitely meet expectations of the sector. While the Kisan Credit Card and the attendant schemes of Insurance are seen as a further dose of encouragement. In the back drop of the macro focus on Indian agricultural, the budget pronouncements have been kind to Indian Coffee, Tea, Copra and Coconut and even Vanaspati oil.

The future looks grim, but it is almost here. Minimum support price is out....well nearly out. The future is one in which the dictum is that this price will not exceed the rate of inflation in any one year. Inputs into agriculture will not be protected and artificially insulated.

MEASURES TO ERADICATE POVERTY & UNEMPLOYMENT

It is evident that growth even agricultural growth, will not lead to a reduction in rural poverty, though it necessary pre-requisite for effective tackling of the rural poverty problem. In order to eradicate poverty from our country, we ought to deploy resources at a higher level and more efficiently to upgrade the production capabilities of those left out of the process of economic development. This will have to be accompanied by structural changes in property relations in rural India. So that the poor get an adequate share of productive assets, including, land, implements, inputs and credit. Eradication of poverty and unemployment from the Indian soil needs reorientation of the national economic policy in the following direction.

1. We must distinguish the potentially productive and non-productive poor. The former will have to be provided with adequate work opportunities and the latter adequate income for the subsistence. Thus our anti-poverty programme should mainly aim at creating large-scale employment opportunities.

2. Massive programme of agricultural development, by extending irrigation facilities wherever possible and evolving dry land farming techniques for unirrigated areas, is a sensible course for the country to adopt. Faster development provide gainful employment to large section of rural population.
3. The problem of rural poverty can be solved on long-term basis if the rural destitutes are provided either with land to cultivate or alternative avenues of employment, assuring fair wages to agricultural labourers and by relieving them of the debt burden which they have inherited from the past.
4. Efforts should also made sufficiently by setting up agro and rural based industries. Most of currently rural unemployed people will get jobs in these industries. Rural industries programmes may be supplemented by massive programmes of rural works aimed at expanding and improving the efficiency of rural infrastructure facilities which will go a long way in increasing the rural productivity.
5. Sufficient investment will have to be made in human capital in the form of education. Education will bring new light to the rural masses and they will be relieved of ignorance, narrow outlook, superstition, laziness etc.,
6. To maximise employment in agriculture, it is also necessary to push forward the implementation of Land re-distribution programme and schemes for the consolidation of holdings and to regulate the growth of farm mechanisation to ensure maximum labour use consistent with optimum land and water utilisation.
7. Those farmers who have uneconomic holdings must be provided with facilities of subsidiary occupations for this purpose. These occupations should naturally be in harmony with agriculture and be land based. These would include rearing of dairy cattle, fruit cultivation, vegetables gardening, poultry farming, bee keeping, mushroom Cultivation, Sericulture, Goat and sheep farming, pigerry etc.,
8. Mere change in technology and mode of production unless accompanied by land reforms will not enable agricultural operation to sustain additional labour. Small scale industrial development in rural areas must be linked with local resources and skill to have an impact on the employment situation.
9. Facilities for provision of Credit, Crucial inputs and marketing will have to be organised. Further farmers and those who are engaged in rural industrialisation should face no problems in the marketing of their agricultural produce and finished goods. They should also be ensured a fair return for their goods.
10. Landless labourers, rural artisons and all those whom the land is unable to sustain, rural industries like handlooms and handicrafts can be viable, and been competitive in export markets, given organisation,

credit supply, small improvement in techniques, design and marketing assistance.

11. One essential element for agricultural growth and poverty eradication is a sound macro-economic environment structure that promotes popular participation and peoples empowerment.
12. Diversification of the economy from predominantly weather depended agricultural economy producing primary goods to a vibrant modem and diversified economy supported by scientific and modern farming techniques capable of exploiting effectively the rich resource base readily available in India.
13. Securing market for products is serious bottleneck for potential farmers. Market information is not readily available and for some, particularly those living in the periphery, they cannot access markets because of poor infrastructure. So proper market and information must be made available to rural community for sustainable agriculture.
14. It is now well known that it not the rate of growth but the composition of growth which determines the pace of the 'spread effect'. The pace is significantly affected by the existence of infrastructure which helps in the 'trickle down' of the benefits of growth. Growth might have created the opportunities but if people are not in a position to respond to these stimuli poverty alleviation cannot take place.

NOTES AND REFERENCES

1. Brundflaud, G.H., et al. (1987), "Our Common Future: Report of the World Commission on Environment and Development", Oxford University Press, Oxford.
2. Bertelmus, P. (1994), "Environment, Growth and Development: The Concepts and Strategies of sustainability", Routledge, London.
3. Myrdal, Gunnar, Asian Drama, Vol. II.
4. Coale, Ansley, J. and Hoover, Edgar, M. (1958), "Population Growth and Economic Development in Low Income Countries" Princeton, New Jersey, Princeton University Press.
5. World Development Report 2000-2001 Attacking Poverty.
6. Gaiha, R. (1991), "Poverty Alleviation Programmes in Rural India: An Assessment", Development and Change, Vol. 22.
7. Balla, Sheila (1995), "Development, Poverty, and Policy: The Haryana Experience", *Economic and Political Weekly,* Vol. 30, Nos. 41 & 42, October 14-21.
8. Parthasarathy, G., (1995), "Public Intervention and Rural Poverty: Case of Non-Sustainable Reduction in Andhra Pradesh", *Economic and Political Weekly,* Vol. 30, Nos. 41 & 42, October 14-21.
9. Technology Information Forecasting and Assessment Council (TIFAC) Report on Food and Agriculture: Technology Vision 2020.

38

ECOLOGICAL FARMING – A FEW TECHNIQUES

K.A. Rasure

INTRODUCTION

There is a growing awareness all over the world of the fact that the natural resources needed to support humanity are indeed, rapidly going in to decline. The supplies of oil and coals are getting shorter as also water and minerals, forest and number of varieties and species of organisms. The spread of world trade, industries and increased consumption of the natural resources by the industrialised countries has certainly been a major causal factor of this depletion. While these are all to the advantages of the developed industrialised countries, the non- industrialised countries suffer most. The overall deterioration in the terms of trade is undermining the economies of industrially under-developed countries, because they are forced to get hold of more and more foreign exchange to pay for industrially manufactured investment goods they need. This deterioration in terms of trade makes itself most strongly felt in the sphere of agriculture and in the countries where the economy is largely dependent on agricultural exports. Because the prices for imported capital goods such as inorganic fertilisers, pesticides and machinaries tend to rise faster than the price paid for the produce itself. The consequential effect is a shattered economy both for the farming community as a whole and for the individual farmers. The local factors such as an under developed infrastructure, the distance between farm and the market, competition with better situated farms, over use of land, and finally the detrimental effects of deforestation on the water balance and soil erosion have further deteriorated the economic conditions of such under developed countries.

Considering the view point stated above, it is not only justified but also necessary for the non industrialised countries to promote a kind of agricultural development which will best serve the emerging need of their respective population growth. Modern agriculture advocated by the industrialised countries is based on the principle of profit maximisation and requires large-scale investment in machines, inorganic fertilisers and pesticides. It specialises in monoculture whereby the diversity found in nature is no longer considered desirable. It is the most artificial system replacing the natural one. Production of high yields necessitates deep ploughing, intensive use of inorganic fertilisers and pesticides, irrigation, weeding, mechanization of sowing and harvesting. These process leads to soil degradation, soil erosion, increased salinity, depletion of organic matter in soil, pollution of air and water. On the contrary, ecological farming attempts not only to guarantee economic security but also to increase yields. This diversity is of course controlled and regulated by man, but the aim of it is to imitate ecological function of the region.

ECOLOGY

The word "ecology" is derived from the words of Greek origin. It is defined as science of interrelation between living organisms and their environment. In India, Hindu scriptures such as Mahabharat mention the importance of food chain and interdependence of organisms for food. However, the modern ecology can be traced back to 17th century A.D. Darwin in 1859 in his book, "origin of species" described in details, the interaction of organisms among themselves and with their environment and their struggle for existence and survival of the fittest. In the 20th century, ecology came in to prominence only after world war-II like the nuclear science and electronics. Today, it is a mature science. "The International Biological Programme of UNESCO has provided further impetus to this science".

The organisms of any community besides interacting among themselves have functional relationship with the environment. This structural and functional system of communities and their environment is termed as ecological system or ecosystem. The study of ecosystem has great bearing on the economy and the welfare of human society. This ecosystems very from location to location according to climate, soil, conditions and topography. Within a region, the composition of natural diversity remains roughly constant or only changes very slowly.

When people came and start clearing the land, working over it, growing crops, harvesting and so on, they thereby create a new ecosystem. This new one may either lead towards a new state of equilibrium with their knowledge of the natural order or else they do not perhaps, partly due to lack of knowledge and partly because of lack of will. In this case, the synergy within the ecosystem is disturbed. A system such as this, doses not have the regulating mechanism found in a naturally diverse system and so, it can not maintain the equilibrium. Eventually, problems arise both regard to maintenance of soil fertility, soil erosion, insect pests and disease syndrome. These are tackled by using chemical substances whilst actual causes are lost sight of.

Recent findings of ecological study have revealed that the nutrient needed for the growth of plants and animals within an ecosystem are constantly passing through practically a self-contained cycle. For example, if the plants are supplied with a moderate dose of nutrients i.e. not too many and in balanced proportion, they can use them effectively and satisfactory yields can be achieved, whilst mineral erosion from the soil is kept to a natural minimum at the same time. A well balanced supply nutrients can be achieved by means of a controlled diversity, because the plant species have different habitat, and they vary considerably in their ability to break down nutrients present in the soil. The soil revives back these in a form which is suitable for other plants to absorb, through green manuring in situ, green leave manuring, compost, dying roots etc. The plants cultivated within an ecosystem have different rate of growth. By following suitable cropping pattern and crop rotation, it is possible to keep the land covered almost constantly. This helps not only to maintain soil fertility but also to protect the soil from water erosion. A controlled diversity of plants can also be used in conjunction with windbreak or farm forestry to help lessen the water and wind erosion.

THE TECHNIQUES

The influencing factors such as climate, soil topography, flora and fauna can vary from location to location, and as such, the farming unit to be organised will also vary. So, a set of framework for ecological farming cannot be formulated, which will suit all farming conditions. Considering the local conditions e.g., social, economic and environmental, farming methods are to be developed for adoption in each particular environmental locations. Among others the following few important techniques should be borne in mind in developing ecological farming.

1. **Diversity of Produce**: It means growing as many kinds of agricultural product as possible suitable to a farming unit. Crops which are needed for farmer's own consumption, should receive priority, Rice Wheat, Pulses, Oil seeds, Vegetables, Fibre and Fruits. On the same piece of land, these crops can be grown following suitable cropping pattern and crop rotations. This ensures (a) proper distribution of work for the farm labourers within the year, (b) continuous farm income because different crops will mature at different times, and (c) economic security to the farmers since the marketing problems is spread over a several products.
2. **Mixed Cropping**: This refers to growing of more than one crop at a time in the same piece of land. This helps in better utilisation of sunlight and soil, brings higher overall yields and farm income and covering of the land for a longer period of time, thus providing protection against soil erosion. Pigeonpea and high land rice and Black gram and Ragi and Pigeonea etc., are few examples. Low land Rice, however, cannot be grown mixed with other crops on account of the way it is cultivated under puddled condition. In case of low land rice, the soil fertility and yield can be improved by following proper crop rotation, green manuring in situ, and use of azolla etc.
3. **Green manuring and mulching**: In between two main crops or during fallow period, various mixture of grasses and leguminous plants can be sown into the soil to provide a covering for it. This will protect

soil erosion. Loose organic material such as dead leaves, grass, straw, pulled weeds left on the ground or even house-hold refuse is suitable for mulching. The rotting of these material increases the fertility of soil, besides protecting soil against water erosion. In order to protect the terraced slopes/undulating high lands, grasses, which form thick roots system, are planned e.g. vetivaria, Zizanoides (vetiveria grass), pennisetum clandestinum etc.

4. **Crop Rotation**: The natural environmental equilibrium of any location cannot be imitated by any farming activity and problems may arise even if a mixed cultivation is carefully practiced. Thus, a well-balanced crop rotation is a useful part of ecological farming. Leguminous, non-leguminous plants, shallow rooted followed by deep-rooted crops, and tuberous plants, with plants, which bear fruits above ground, are to be included in the crop rotation. It helps checking spread of diseases and pests, which are found with monoculture.

5. **Growing Range of Varieties**: In ecological farming, it is important that instead of growing high yielding varieties which have particular nutritional requirements and are susceptible to insect pets and diseases, yield-stable varieties which will give optimum yield in less favorable conditions, should be planted. Here, too, diversity helps checking the spread of decease and pests.

6. **Compost**: Supply of plants nutrients should be made through or ganic manure, compost etc., and use of chemical fertilisers should be minimised, in ecological farming. Compost may be briefly described as decayed organic matter. Vegetable waste such as straw, grass, dead leaves, sawdust, as well as animal residues as dung, urine, dried blood, bones, hairs etc., can be used for compost making. Water, air and alkaline materials like crushed lime stone, wood ash and earth should be used to ensure effective rotting process in the compost pit. Compost cantinas almost all the plant nutrients necessary for plant growth. It plays an important role in soil humus formation and improving physical, chemical and biological properties of the soil.

7. **Integrated Pests Management**: If these above techniques are fully developed and practised, there are less chances that the pests and diseases will take on serious proportions. The integrated pest management includes maintenance of proper soil health, field sanitation, planting resistant varieties, biological control and least use of chemical substances if required.

8. **Mixed Farming**: Where it is possible to keep dairy animals, it is desirable to include them in crop husbandry. Mixed farming not only helps utilise the farm by-products but also helps producing high animal fertilisers such as farm-yard manure and compost. Good matured compost has the effect of regulating and stimulating both the formation of humus and the fertility of the soil.

9. **Integration of farm forestry**: A natural, location-oriented ecosystem is most widely achieved in wooded area. Reforestation is beneficial in areas where there is no forest. Farm forestry can help revitalising the soil and regulating the water balance. It can also supply firewood and timber for construction of farm house and farm implements. It is desirable to included farm forestry with arable farming and to discourage monoculture.

CONCLUSION

Small farmers constitute the bulk of farming community. They seek improvement of their agricultural methods. In many cases the traditional farming systems are not found suitable. The modern high cost farming methods do not offer a realistic alternative for these farmers. Ecologically oriented farming is needed to establish high and lasting soil productivity, basis of sustained agricultural production and independent development.

39

SUSTAINABILITY OF AGRICULTURE IN DEVELOPING COUNTRIES

R. Sundareshan

SETTING

Agriculture has been the economic activity over years in many countries, which is most essential to human survival and well-being. It has also been the economic sector which most affects, and is most dependent on, the natural environment. Agriculture in developing countries has the characteristics such as: farms are small, and they are threatened with the degradation of natural resources and the environment. It is concluded that the root causes of environmental degradation are social an institutional in nature. Therefore, measures to address the problem will involve an adjustment of polices and institutional structures. The world Trade Agreement of 1994 brought agriculture for the first time in world trade history within its policy frame work. This is brought about by the singing of Uruguay round of trade negotiations.

GENESIS OF GATT/WTO

Seed of the GATT was sown in 1947 in Geneva with total member strength of 23 countries to issue the objective of free trade in order to encourage growth and development of all member countries. GATT which is otherwise called as the programme of globalisation is formed based upon the concept of considering the whole world as the global village. The main objective of GATT was to ensure competition in commodity trade through reduction in tariff restrictions on imports imposed by member countries. The Uruguay round of talks which was technically called as Multilateral Trade Negotiation Conference (MTN) was launched at Punta deleste in 1986.

Since 1948, World Trade had undergone a lot of structural changes over the last four decades in many developed countries. The share of agriculture in world trade had declined from 46 per cent in 1950 to 13 per cent in 1987. The structure of employment and contribution of various sectors to GDP of developed countries and undergone a qualitative changes and comparative cost advantage in many commodities had moved in favour of developed countries. Developed countries under the leadership of USA take the initiative of bringing service sectors into trade negotiations. Finally the Uruguay Round contain the mandate to have negotiations in 15 areas of which 14 areas related to trade in commodities and the one on trade in services. A part from tariff, non-tariff barriers and antidumping measures etc., new areas viz., Trade Related Intellectual property rights (TRIPS) Trade Related investment measures (TRIMS) and trade in services were included for the first time for negotiations.

Because of certain critical issues, the agreement suffered a blockade. Mr. Arthur Dunkel, Director general of GATT submitted the report, popularly known as 'Dunkel Proposals' on 20th December, 1991. On the basis of this report, in 1993, one more conference was held at Geneva, where the report was finalized. The final agreement was made in 1994 and 117 counties signed the agreement including India, at Murakash in Morocco and World Trade Organisation (WTO) was constituted.

WHAT IS SUSTAINABLE AGRICULTURE?

It is relatively easy to describe goals for a more sustainable agriculture, things become much more problematic when it comes to attempts to define sustainability. Andrew Campbell (1944 a) says that 'attempts to define sustainability miss the point that, like beauty sustainability is in the eye of the beholder..... .it is inevitable that assessments of relative sustainability are socially constructed, which is why there are many definitions. FAO has defined sustainability as, "...The management and conservation of the natural resource base, and the orientation of technological and institutional change in such a manner as ensure the attainment and combined satisfaction of human needs for present and future generations. Such sustainable development (in agriculture, forestry, and fisheries sectors) conserves land, water, plant and animal genetic resources and is environmentally non-degrading, technically appropriate, eco-nomically viable and socially acceptable". (FAO) 1991 a)

SUSTAINABLE AGRICULTURE IN DEVELOPING COUNTRIES

Agriculture in developing countries has the following two important characteristics. Farms are generally small, and most countries are suffering from a degradation of natural resources and the eco system. Sustainable agricultural practices which are suitable for developing countries must take into consideration the fact that population growth rate is high, that caused increase in food demand at a rate of 3-5% per year (Dent 1989). This rapid increase in population has caused farmers to expand into marginal areas and make more intensive use of marginal land, leading to degradation of land and water and would result in poverty, malnutrition and poor standard of living. A developing country's like India is surplus in food grains and has over 45 million tones of wheat and rice in the godowns, but in the same time over 250 million children, women and men go to bed partially hungry everyday. From being food-secure, these countries needs to become livelihood-secure. An that is the biggest challenge facing developing countries today.

Table 1
Projected Household Demand for Food in India at 7% Income Growth

Commodity	*Annual Household Demand (Million Metric Tonnes)*				
	1991	*1995*	*2000*	*2010*	*2020*
Food grains	168.3	185.1	208.6	266.4	343.0
Milk	48.8	62.0	83.8	153.1	271.0
Edible Oil	4.3	5.1	6.3	9.4	13.0
Vegetables	56.0	65.7	80.0	117.2	168.0
Fruits	12.5	16.1	22.2	42.9	81.0
Meat, Fish	3.4	4.4	6.2	12.7	27.0
Sugar	9.6	10.9	12.8	17.3	22.0

Source: Technology and Information Forecasting and Assessment Council; Food and Agriculture, Technology Vision 2020.

Internal and external factors are threatening the future of agriculture and the food security of the people of developing countries. Among the internal factors, the most important one relates to the growing damage to the ecological foundation essential for sustained advances in biological productivity. Among the external factors the most regions are the unequal trade relationship, increasing proprietary control over science, and potential changes in climate and sea level. These threats need attention.

IMPACT OF WTO ON SUSTAINABILITY OF AGRICULTURE IN DEVELOPING COUNTRIES

The principles and polices of the WTO are based on Ricardo's concept of comparative advantage. The four major elements of the World Trade Agreement in the field of agriculture are:

- Market access
- Domestic support
- Export subsidies
- Trade-Related intellectual Property Rights

MARKET ACCESS

Under the market access commitment, all number countries of WTO are requested to:

(i) Replace all types of non-tariff barriers with tariff barriers and

(ii) Reduce the level of tariff under a time bound programme.

As per item (i) it could be observed that under the Agreement on Agriculture of the WTO, agricultural sector ha about 800 items, the imports of which are subject to Quantitative Restrictions (QRs).

In the recent settlement, India had made two demands to safe guard the farmers. They are:

The first is call for a safe guard mechanism for all developing countries along with a provision for imposition of Quantitative Restrictions under specified circumstances in the event of a surge in imports or a decline in prices. The second is a demand, that as special and differential measure, the developing country members should be allowed to maintain appropriate levels of tariff bindings keeping in the mind their developmental needs and the high distributions in the international markets.

QUANTITY RESTRICTIONS (QRs)

A Weakening Gap

As on March 31, 2001 QRs will be removed on all commodities, including agricultural commodities. Market access place a demand on quality and efficiency of agricultural operations of the developing countries. Since, developing countries agree for retention of QRs on agricultural imports on grounds of food security, in domestic policy, they are rushing recklessly into implementation of the QRs removal. The gap between their negotiating position in the WTO and domestic position in the WTO and domestic policy weakens their stand in the WTO and is also a threat to the farmers.

The process of lifting of QRs is not a new one and has been going on since the mid 1990's. Therefore the abolition of QRs on 75 products on April 1,2001 is not a new phenomenon-except for the fact that the last list comprises largely "sensitive products" on which import restrictions have been maintained to the very end.

Table 2 gives a comparison of imports before and after the abolition of QRs with respect to India. Two kinds of comparisons were made.

Table 2
Effect of Removal of Quantity Restrictions (Qrs) Before and After

		Category I		*Category II*	
		Products on which QRs were lifted on April 1, 1999		*Products on which QRs were lifted on April 1, 2000*	
		1998-99 (QR era)	*1999-2000 (Post Q Rera)*	*1999-2000 z(QR era)*	*2000-2001 (Post Qrera)*
1.	Value of imports (Rs. Crores)	11,802.00	12,821.00	5,151.56	2,675.00
2.	As % of total imports	6.6%	6.9%	2.6%	2.3%
3.	Number of products imported	554	622	487	474
4.	Value of Agricultural imports (Rs. Crores)	8,123.00	8,619.00	510.00	405.00

(i) 894 products were feed from QRs on Aril 1, 1999 referred to here as category I. Import of these products in 1998-99 (QR era) are compared with their imports in 1999-2000 (post-QR era).

(ii) 714 products were put on the free list on April 1,2000 (Category II). The comparison here is between imports of these products in 1999-2000 and in the first six months of 2000-2001.

The results of analysis denote,

- Of imports on the category I list, there is a slight increase. On the category II list, there was even a small fall in the number of products imported after the QRs were removed.

Since large number of agricultural commodities made up in category I products and figured very little in category II, their imports too reflected the same pattern. The growth of category I agricultural imports was very minimum increase of 6 per cent only in 1999-2000 and inferred that currently there is no sudden jump in imports. One will have to wait and see what will happen, when the QRs on the last and most sensitive products are removed on April, 1.

TARIFF REDUCTIONS

Under item (ii), i.e., tariff reduction, the levels are to be reduced by 24% in the developing countries and 36% in case of developed countries and the corresponding period during which there reduction are to be taken up are six years for developed countries and 10 years for developing countries. Besides these commitments, this measure also calls for maintaining current access opportunities and establishment of minimum access tariff quota is to be established at reduced 'tariff quota'. The minimum access tariff quota is to be established at reduced tariff rates for those basic products where are current market access is less than 3% of domestic consumption. (Later considered to rise gradually to 5% of domestic consumption).

The policy was developed with care over all the member countries, the reality is different. The AOA has allowed industrialized countries to maintain high levels of tariffs. As India's paper for the AOA review points out, the average tariffs in OECD (Organisation for Economic Co-operation and Development) countries in 1993 were 214% for wheat, 197% for barley and 154% for maize. In certain commodities the tariff peaks reach 350% and above. In the European Union, the out-of-quota tariff for bananas is 180%, in Japan these tariff range between 460 and 600% for dried beans, peas and lentils, and in the U.S., groundnuts have a tariff of 164%. Recently, Japan has levied a tariff of about 1000% on rice. In such a case, the developing countries need to raise the duties at par with industrialized countries imports duties. This is allowed under the WTO.

The removal of QRs and the low level duties had led to the collapse of farm prices, especially food grains, the most significant aspect of the agricultural crisis, in the developing countries. Prices for rice plunged from Rs. 968 per quintal in 1998, to Rs. 862 1999 and further declined to Rs. 805 in 2000. Wheat price plunged from Rs. 759 in 1998 to Rs. 740 in 1999 and Rs. 605 in 2000. Akhar dal prices fell from Rs. 2,380 in 1998 to Rs. 1925 in 1999 and Rs. 1400 in 2000. Hence, the developing countries like

India should take care of these aspects while preparing domestic polices and to protect farming community's well being.

Tariffs are governed by certain binding commitments that India made before the WTO at the time of its accession to the WTO regime. About 1,429 product lines that have been decontrolled or stand on the verge of being decontrolled, India has infinite flexibility to impose tariffs of its choice in no fewer than 863. This means that India is not bound by any commitments on the highest level of tariffs that can be improved on the imports of these products. In a situation of distress for domestic producers, India would have a wide measure of latitude to increase protective tariffs. This would have wide measure of latitude to increase protective tariffs. This would have provided some measure of reassurance for domestic agriculturists.

AGGREGATE MEASURE OF SUPPORT

AMS is the annual aggregate value of market price support, non-exempt direct payment and any other subsidy not exempted from the reduction commitment expressed in monetary terms. Polices, which have been excluded from reduction commitments, are government spending on research, disease control, infrastructure and food security. These include structural adjustment assistance, direct payments under environmental programmes and regional assistance programmes.

In product specific AMS, the market price support are calculated to find th difference between fixed external reference (FOB unit value in a net exporting country and CIF unit value a net importing country) and the administered price (domestic support price) multiplied by quantity eligible to receive the administered price. However there is not enough clarity in the final act whether the quantity eligible to receive administered price is the total quantity of production or only the marketed surplus.

The non-exemption of direct payments are to be computed by using the gap between external reference price and administered price of \r by using budgetary outlays. Input subsides and other polices like marketing cost reduction measures are to be determined by using budgetary outlays. If the budgetary out lays do not reflect the full extent of subsidy, the subsidy has to be worked out by using the difference between the subsidized price of input and a representative market price multiplied by the quantity used of particular input.

As per the GATT/WTO agreement if the product specific and non-product specific AMS doesn't exceed 10% of the total value of agricultural product in the case of developing countries, the country is not required to reduce the product specific and non-product specific support. However, if the AMS exceeds 10% of the total value of agricultural production, it is required to be reduced by 13.3% in the case of developing countries of the value, that doesn't qualify for exemption during the implementation period.

Detailed guidelines have been evolved to compute the "aggregate measure of support" (AMS) that Agriculture enjoys in relation to the base of the WTO-agreement, which is the period 1986-88. And developing upon the stage of development, every country is required to scale down the AMS to an agreed quantum, as it affords. A developing country like India is in the comfort zone as far as these norms are concerned, since it has claimed before the WTO that agriculture in the country, far from enjoying any positive subsidies, is actually negatively protected by the administrated price regime.

DOMESTIC SUPPORT

The industrialized countries have introduced or number of safeguards such as the "Blue" and "Green" Boxes in the WTO agreement that provided them with cover for providing 'domestic supports' for their farmers. The provisions in the green box include policies that provided services for benefits to the agriculture rural community, stock holding for food security, domestic food aid, investment subsidies and agricultural input subsidies for low-income and resource poor families.

The provisions in the Blue box include direct payments to farmers under production limiting programs. For example, the United operates an extensive scheme to match production and projected market demand. The farmers are paid not to produce a crop, yet the payment is not called a subsidy. This has helped ensure a steady and assured income to the farmers. Unfortunately, farm families in the developing countries are expressing all the negative impacts of the agreement, mostly due to their own inaction or lack of timely action.

Former U.S. President Mr. Bill Clinton in his recent address at Warwick, U.K., stressed the need to ensure that trade is carried out with a human face and that is becomes and instrument of poverty eradication. Majority of population in developing countries desire their livelihoods from crop production, animal husbandry, forestry, inland and marine fisheries, agro processing and agri-business. Generally they belong to the small and marginal farmers category. The reality is that the post-harvest and sanitary and phyto-sanitary infrastructure is very poor and massive investments in this area will be essential, if these countries are to ensure a level playing field in the field of external trade in agriculture.

Developing countries had under expectations of the developed countries reducing subsidies, making the products more competitive in the external market. But that has not happened because developed countries continued their subsidies in different ways. In developed countries, the farming community constitutes only 2 to 3% of the total population, it is vital for the survival of the respective countries because the ecology is just not seen as a food producing machine but as vital activity for the good of the environment. It is one with multifunction such as clearing the air, absorbing carbon-dioxide and sustaining ecology. Thus, agriculture has a multi-functionality role in terms of the environment in the developed countries. Whereas in India and other. Developed countries, its multi-functionality role is greater as it not only a means of producing food for the urban people, but a means of livelihood and a way of life for millions of people.

EXPORT COMPETITION

In the area of export competition, GATT/WTO agreements calls for reducing direct subsidies to a level of 36% below 1986-88 level in the case of developed countries over a period of six years. The quantity of subsidized exports by the developed countries is to b reduced by 21% during the same period. In, the case of developing countries direct export subsidies are to be reduced by 24% and the quantity of subsidized exports is to be reduced by 14%. These reductions are to be carried out over a period of 10 years as against ix years in case of developed countries.

Export subsides included in reduction commitments are:

(i) Direct subsidies paid by the government any other agency

(ii) Payments that are made from the proceeds of levy imposed on agricultural products

(iii) Subsidies given to reduce the costs of marketing including internal handling, processing, international transport and freight subsidy on export shipments.

Subsidies available to exporters of agricultural commodities are in the form of:

(i) Exemption of profits from export sales from the income tax; and

(ii) Subsidies on costs of freight on export shipment of certain products like fruits, vegetables and floricultural commodities.

In contrast, the total farm support increased by 8% to $ 363 billion in 1998 in Organisation for Economic Co-operation and Development countries. Experience during the last six years has shown that agricultural exports would not be improved, otherwise international process of several Agricultural produce have been increased. During 1999-2000, the price for inter wheat was down by $ 20 a tonne; and prices for coffee declined by 28 cents a pound in January 1999 and by 72 cents in September-now the price of cotton all the lowest since 1993. Similarly the prices of cotton are the lowest in the last 13 years. The main thing is that while multinational companies and traders control most of the trade, they are not the produces. Competing in the world market with these kinds of companies will be a hard game for the Developing countries.

TRADE RELATED INTELLECTUAL PROPERTY RIGHTS (TRIPS)

The philosophy underlying the existing legislations and jurisprudence on IPR matters in developing countries covering copy rights, trademarks, patent and industrial design is that the protection of IPR is desirable for the encouragement of beneficial creative within the country. The most talked about agreement under GATT is TRIPS. TRIPS is one of the major trade issuers relating to agriculture and it is one which will have significant impaction sustainability of agriculture in developing countries.

TRIPS comprise guidelines for protection of seven categories of Intellectual properties viz.,

(1) Patent (2) Trade magic (3) Copy Right (4) Industrial Design (5) Design of Integrated Circuits (6) Geographical indication (7) Protection of Undisclosed Information. Except in patents and geographical indications polices, laws and regulations and administrative and judicial system of developing countries are almost similar to the rest of the world, and patent law and geographical indications have a direct effect on agriculture.

According to WTO agreement, developing countries are required to bring their laws and regulations in conformity with obligation under the agreement. The private sector agricultural research is largely promoted and funded by input producers with profit motives without considering social, economic or ecological impact at large. In the case of seed industry, as the demand for hybrid seeds increasing major quantity is produced by the private seed industry, the demand for protection is also increasing. Private seed companies concentrate on hybrids (terminate gene, verminator gene) which the farmers have to renew every year instead of composites, which need to be replaced less frequently. The private seed companies, mostly multinational companies can create a monopoly situation and earn enormous profits. On the positive side, such a system will make available improved and better quality seeds in the market, and major beneficiary will be even agricultural research institutions and most appropriately the farmers.

The small seed industries may not be able to compete with multinational companies. So, These industries should either increase their competitions or to merge or form joint venture with the foreign companies to with stand in the market. The research and development activities undertaken by the multinational companies also bring in foreign exchange, improved techniques and quality products. But we should be cautions about the increasing costs of hybrid seeds.

The recent issues about patenting show clearly the challenges imposed by TRIPS. This phenomenon of bio-piracy in the patenting of indigenous knowledge as invention by western scientists and individuals should be dealt with seriously. Geographically indications are another major area of concern. This provides protection to goods, which originate from a particular area. But these is no obligation to protect goods, which are not protected in the country of origin. This should be taken case by the governments of developing countries, through legislations or forming gene banks.

Unlike the general perception of many the TRIPS agreement provides both in letter and spirit, for considerable maneuvering to make it developing countries friendly. For example under Article 8.1, it states that "Members may, in formulating or amending their national laws and regulations, adopt measures necessary to protect public health and nutrition, and promote the public interest in technological development, provided that such measures are consistent with the provisions under this agreements".

LEGISLATION TO PROTECTED PLANT VARIETIES

The *sui generis* system as developed by the Indian Government provided rights for the three roles of farmers-as cultivator, breeders and conserver of genetic material. As a cultivator he has plant–back rights–that is to keep seeds for the next season and enter into a limited sale without using a brand name. As a breeder, when a farmer develops a new variety of crop he also has right of sui generis protection.

The conserver–and who–adds value to crops as in the case of medicines–has to get recognition and reward under the National gene fund. Developing countries should insist on the incorporation of ethics and equality provisions of the convention on biological diversity. Care should be taken so that there is harmony between the commitments made by governments industrialized and developing ones–in the Convention on Biological Diversity and WTO agreements.

CONCLUSIONS AND SUGGESTIONS

The visionary farm scientist and food policy expert Dr. M.S. Swaminathan warns: "If our agriculture goes wrong, nothing else in our economy and social falsie will have a chance to right". This warning is not just to India but for all the Developing countries since Agriculture farms the back bone of most of the developing countries. In the globalisation era, the sustainability of agriculture takes that much importance.

1. The market access commitment, due to competitiveness, there is every chance of producing quality Agricultural commodities in the developing countries and thereby fetch higher prices for these commodities in the international market. But due to the lack of cold storage systems and other improved technologies, they may not be able to compete with developed countries. In such a case to safe guard the sustainability of agriculture, during negotiations with WTO, it should be insisted for continuing the QRs on agricultural commodities. The need of the hour is not just food security but to keep in pace with the population explosion, an insistence for livelihood security is very important.
2. Increased industrial competition and reduced protection were expected the improve terms of trade of agriculture, and further benefits in these direction were expected to follow from the agreement on agriculture in the WTO which mandated subsidy cuts in developed countries.

   At the same time, greater reliance on free markets was expected to deliver higher incomes and greater efficiency to the farm sector while enabling subsidies to be reduced.
3. If there is a lower level of import protection to agriculture than to industry, this will result in increase in agricultural prices and this will end the negative subsidies. From a long term vision, this shows a positive sign for the sustainability of agriculture in the developing countries. Although agricultural prices have weakened in the last two years there is relatively high level of production of agricultural commodities. There has been a decline since, but even so the terms of trade during the last two years have continued to be amongst the highest in the last 25 years.

4. The removal of QRs pose a threat to the developing countries, the allowance regarding the alternation that shall be made in the import duties gives some relief. If proper policy measures are adopted, this could be well tackled.
5. To avoid the exploitation of developed countries over developing countries, WTO should take an effective monitoring and ensure that the member countries strictly adhere to the rules and regulations in various trade aspects. For this the developing countries should so that advanced economics do not in a some what unilateral manner decide product standards among themselves and force the developing countries to meet these expensive standards.
6. Regarding TRIPS, developing countries should insist on the incorporating ethics and equality provisions of the Convention on Biological Diversity. The *sui generis* system is a good one to enjoy the real benefits of TRIPS by the Agriculturists in the developing countries.
7. In order to encourage agricultural exports, the subsidies are to be continued atleast until the developing countries achieve self sustainability in export.

REFERENCES

A.P.J. Abdul Kalam (1998), India 2020.

Jules N. Pretty (1995), Regenerating Agriculture–Polices and Practice for Sustainability and Self Reliance.

C. Ram Manohar Reddy and V.S. Sambandam (2001), "Limited Growth in Post–QR Imports". The Hindu, March 8.

M.D. Nair (2001), "Amendments to Patents Act–options under TRIPS", The Hindu, March 8.

M.S. Swaminathan (2001), "Livelihood Security", Frontline, Feb. 16.

Rajech Mehta (2000), "Removal of QRS and Impact on India's Import", Economic and Political Weekly, May 6.

Ruddar Datt and K.P.M. Sundaram, Indian Economy.

Vaidyanathan A. (2000), "India's Agricultural development Policy", Economic and Political Weekly, May 13.

Vandana Shiva (2001), "India's Agricultural Policy, Budget and WTO", The Hindu, Mar 6.

"The farming Crisis, The Balance Sheet Optimism", Frontline, Feb. 2, 2001.

Survey of Indian Agriculture 2000, The Hindu.

Impact of WTO on Tamilnadu Agriculture and Agro based products–Center for Agricultural and Rural Development Studies (CARDS), TNAU, Coimbatore–3, 2000.

Sustainable Agriculture for the Asian and Pacific Region, "Food and Fertilizer Technology Center for the Asian and Pacific Region" Book Series No. 44.

40

CROP DIVERSIFICATION AND SOIL & WATER MANAGEMENT IN ORISSA

Jagannath Lenka and Minati Mallick

INTRODUCTION

Natural resources determine the course of economic development. (W.A. Lewis). Countries endowed with natural resources are expected to have a natural advantage in economic growth. But it is observed that the natural resource rich countries on an average have grown less rapidly than the countries poor in natural resources.

A judicious use of natural resources in a sustainable manner is the basic requirement to set the wheel of economic progress in motion. Both economic use of exhaustible natural resources and conservation of renewable natural resources are equally important for the purpose. But growth being at the center stage of the policy agenda of almost all countries in the world has led to fast depletion of natural resources. The natural resources as a form of capital, if depleted, either has to be regenerated or substituted if countries are to maintain or expand their asset base and thus ensure a constant or rising trend of economic growth (Dasgupta and Heal, 1979).

Land and water constitute two important renewable resources extensively used in agricultural sector. A change in climate, non-economic use of land and water, cultivation of highly water consuming crops have resulted in a fall in fertility of soil, soil erosion and depletion of water table. The conservation of these renewable resources is of utmost importance for viability and sustainability of farm enterprises.

B.B. Vohra of National Commission of Environmental Planning in his Sardar Patel Memorial Lecture has pleaded for a reduction in the cultivable area on the ground that increase in economic efficiency thereby can produce sufficient food grains to support countries population.

The diversification of agriculture is an alternate way not only for the regeneration and conservation of land but also for enhancing its productivity. Moreover, it may result in conserving the most important resource that is water. National Agricultural policy (NAP), 2001, emphasised the need for efficient use of soil, water, bio-resources to meet the increasing requirement at home and face the competition in world market efficiently.

With this backdrop the present paper makes a modest attempt to analyse the trends and determinants of agricultural diversification in the state of Orissa. The paper is organized as follows. The second section presents the post reform challenges. Data and Methodology are given in section–III. Results and discussion are presented in Section-IV. The concluding section contains summary of the findings and policy options.

POST-REFORM CHALLENGES

A significant change in Indian Agriculture has taken place in post reform era due to liberalization and globalization through increasing export orientation and market access to others. To cater to these requirements the National Agricultural policy (NAP), 2001 emphasized the need for efficient use of soil, water, bio-resources, fertilizers etc. The NAP also visualized technological transformation of Agricultures so as to cultivate location specific and economically viable varieties of crops, use of biotechnology, strengthening the linkage between research and extension activities, provision of infrastructural facilities, and legalisation of tenancy to promote contract farming system.

In the context of the policy changes after economic reforms, the change in the production pattern in agricultural sector and its participation in world trade is noteworthy. Agricultural export hovered around 6 billion dollars with a declining share of cereals overtime. Though the production of food grain has increased from 176 MT from 1990-91 to 212MT in 2001-02 the area under food grain has shown a declining trend but there has not been much shift in the area under food grains to nonfood grains.

As compared to other states Orissa has not shown remarkable progress in agrarian development in post reform period. The State shows fluctuating and declining agricultural performance that can be attributed to natural calamities. Cereals still remain the dominant crops with little diversification of commercial crops confined to a few districts of Orissa.

Orissa predominantly an agrarian state with 84% of its total population having agriculture as their mainstay of life. Agricultural labour accounts for 73% of the economically active labour force of the state. The cropping pattern in Orissa both in rainfed and irrigated situation is rice based that is a highly water consuming crop. The rice invites a low market price lowering the net return over cost. Orissa can divert 10 lac hectares of land out of 45 lac hectares presently under rice towards the cultivation of Cash crops without hampering the total supply of rice in the state. This diversification can raise farm income checks distress sale of the food grain and provide employment in farm sector when the crop is affected by natural calamities. The farmers definitely prefer that combination of crops from which they can derive maximum possible net revenue at least possible risk that will sustain their income on the long run.

DATA AND METHODOLOGY

The present study makes use of secondary data drawn from statistical Abstracts of Orissa, Economic Survey and Orissa Agricultural statistics. The analysis is based on the data at two points of time 1992-93 and 2002-03.

A change in the agricultural pattern is measured through the variation in the proportion of area under different categories of crops in the gross cropped area. For measurement of agricultural diversification, Herfindahl Index (HI) is used in the present study.

$$HI = \sum_{i=1}^{n} P_i^2$$

Where Pi refers to the proportion of area under i[th] category of crops in the Gross Cropped Area (GCA).It takes the value between 0 and 1, when HI = 0, it indicates perfect diversification whereas HI = 1 presents perfect concentration.

The determinants of crop diversification is analysed using the linear regression model

$$Y = \alpha + \beta_1 X_1 + \beta_2 X_2 + \beta_3 X_3 + \beta_4 X_4 + \beta_5 X_5 + \beta_6 X_6 + U$$

Where

Y = Herfindahl Index of crop diversification.

X_1 = Fertiliser consumption per hectare of GCA

X_2 = Proportion of irrigated area in the GCA

X_3 = Rainfall in millimeters

X_4 = GCA per tractor / trailor.

X_5 = Crop loan

X_6 = GCA per Regional cooperative marketing Societies.

U = Stochastic disturbance

In order to study the impact of these factors as crop diversification, a cross section regression model is estimated taking HI of crop diversification as dependent variable. The positive sign of the coefficient indicates declining diversification and negative sign indicates the increasing diversification.

RESULTS AND DISCUSSION

Cropping Pattern

A number of crops are grown in different districts of the state depending upon the local agro climatic conditions. For simplicity of the analysis these crops have been divided into nine broad categories such as Cereals, Pulses, Oilseeds, Vegetables, Sugarcane, Spices, Tobacco, Fibres and Fruits. In order to understand the changes in the cropping pattern that have been taking place at the state level from 1992-93 to 2002-03,the proportions of the area under crops grown to gross cropped area have been estimated and presented in Table 1.

Table 1
Area under Different Crops as proportion of Gross Cropped Area

Crops	*Year*		
	1993-94	*2002-03*	*% Change*
Cereals	0.6400	0.7179	12.16
Pulses	0.1347	0.0996	-26.04
Oil Seeds	0.0693	0.0458	-33.92
Vegetables	0.1202	0.0975	-18.90
Sugar cane	0.0019	0.0024	25.97
Spices	0.0244	0.0243	-0.56
Tobacco	0.0014	0.0006	-54.59
Fibers	0.0080	0.0119	48.24
Fruits	0.0186	0.0328	76.34

It is observed that the major crops grown in the state belong to the Cereals. It covered an area of 64 per cent in the year 1992-93 and increased to 71.79 per cent in the year 2002-03 registering an increase of 12.16 per cent during the study period. The same trend is also observed in almost all the districts of the state.

The next major crops in the state are Pulses and Vegetables. Area under these crops as a proportion of GCA stand in the order of 13.47 per cent and 12.02 per cent respectively in the year 1993-94.It is found that area under Pulses as a proportion of GCA has declined by 26.04 per cent in the year 2002-03 while it has declined by 18.9 per cent in the case of vegetables.

In the case of Oilseeds this figure has also declined from 6.93 per cent in the year 1993-94 to 4.58 per cent in the year 2002-03.

Area under Sugarcane, Spices, Tobacco, Fibres and Fruits constitute only around 3 per cent of the GCA in the state. It is observed that the proportion of area under sugarcane, fibers and fruits have increased while that under spices and tobacco have declined during the period of study.

Though there has some changes in the cropped pattern in the state during the period 1992-93 to 2002-03,the position of the cereals has increased. The area under the commercial crops like Oilseeds and Vegetables, Spices and Tobacco has declined which is a matter of concern particularly after economic reforms. A clear shift in favour of the cultivation of Fruits, sugarcane and fibers is observed during this period.

Crop Diversification Indices

Herfindahl indices are estimated to show the crop diversification across the districts of Orissa.These estimates are shown in Table 2. A cursory glance at the HI values for the state as a whole (0.4647 for the year 1993-94 and 0.5456 for the year 2002-03) reveals the fact that there is crop

concentration in stead of crop diversification during the study period. It is supported by the observation in the previous section that there has been increasing area under cereals.

Table 2
Herfundahl Index of Crop Diversity in Various Districts of Orissa

District	*1993-94*	*2002-03*	*% Change*
Angul	0.3715	0.4122	10.95
Balsore	0.5471	0.6821	24.67
Baragarh	0.2227	0.7366	230.68
Bhadrak	0.7299	0.7024	-3.76
Bolangir	0.4125	0.5358	29.89
Boudh	0.4754	0.5257	10.59
Cuttack	0.3628	0.4681	29.00
Deogarh	0.5286	0.4086	-22.69
Dhenkanal	0.4347	0.6611	52.07
Gajapati	0.4100	0.4031	-1.69
Ganjam	0.4304	0.5167	20.03
Jagatsinghpur	0.4357	0.4426	1.57
Jajpur	0.4665	0.5125	9.84
Jharsuguda	0.5440	0.6659	22.41
Kalahandi	0.4042	0.5789	43.23
Kendhamal	0.2649	0.3274	23.60
Kendrapara	0.5437	0.5148	-5.32
Keonjhar	0.5068	0.2618	-48.35
Khurda	0.5010	0.5270	5.19
Koraput	0.4060	0.4795	18.08
Malkangiri	0.4639	0.5742	23.78
Mayurbhanj	0.5268	0.6683	26.87
Nabarangpur	0.5402	0.6173	14.28
Nayagarh	0.4503	0.4997	10.97
Nuapara	0.4304	0.6030	40.08
Puri	0.5768	0.6560	13.72
Rayagada	0.3060	0.3303	7.95
Sambalpur	0.5533	0.6833	23.50
Sonpur	0.5853	0.7280	24.38
Sundargarh	0.5101	0.6222	21.97
ORISSA	0.4647	0.5450	17.27

At the district level, crop diversification is found to be highest in Keonjhar followed by Deogarh, Kendrapada, Bhadrak and Gajapati.The rest of the districts are observed to have concentration of crops and it is highest in Baragarh followed by Dhenkanal. Thus it is very difficult to attribute any geographical and regional factors responsible for such diversification or concentration of crops across the districts of Orissa.

Factors Influencing Crop Diversification

Several factors influence the pace of agriculture diversification from staple food to high value commercial crops. Crop diversification is generally possible when the conditions required for various crops exists for growing varieties of crops. The required conditions include water resources, suitable soil, moisture and climatic conditions, use of seeds, fertilizers and pesticides, agricultural equipments, availability of finance, insurance provisions and facilities for marketing of the produce. The present study examined these factors of crop diversification and it is found that the most significant factors influencing crop diversification are fertilizer consumption, number of tractors and trailors, regional cooperative marketing societies, rainfall, irrigation and finance. The estimated value of the parameters of the regression for 1992-93 and 2002-03 are presented in Tables 3 and 4 respectively.

Table 3
Estimated Value of the Parameters of Regression model – 1993-94

N = 30	*Adjusted R^2 = 0.6732*		*Standard Error of Estimate = 0.053*
Variables	β	*Standard error*	*t*
Constant	0.0358	0.1843	0.1942
Fertiliser consumption per hectare	-0.3295	0.5134	-0.6418
Proportion of irrigated area	-0.0026***	0.0014	-1.8571***
Rainfall	0.0002**	9.981E-05	2.1683**
Gross Cropped Area per tractor/ trailors	-0.3215*	0.1052	-3.0561*
Crop loans	-0.2124**	0.0829	-2.562**
Gross Cropped Area per RCMS	-0.0031	0.0056	-0.5536

*Significant at 1 per cent level
**Significant at 5 per cent level.
***Significant at 10 per cent level.

Table 4
Estimated Value of the Parameters of Regression Model–2002-03

N = 30	*Adjusted R^2 = 0.5329*		*Standard Error of Estimate = 0.053*
Variables	β	*Standard error*	*t*
Constant	0.2119	0.0899	2.3575
Fertiliser consumption per hectare	-0.5124	0.4615	-1.1103
Proportion of irrigated Area	-0.0043**	0.0016	-2.6875**
Ranifall	0.0014	0.0123	0.1138
Gross Cropped Area per tractor/ trailor	-0.8751**	0.4128	-2.1199**
Crop Loans	-0.0027***	0.0015	-1.8000***
Gross Cropped Area per RCMS	-0.0021**	0.0008	-2.5547**

* Significant at 1 per cent level
** Significant at 5 per cent level.
*** Significant at 10 per cent level.

The model estimated for 1992-93 reveals that the factors retained in the model together explain 67.32 per cent of the variation in the crop diversification index.

Fertilizer consumption, proportion of irrigated area, GCA per tractors/ trailor, crop loansand GCA per RCMS have positive impacts on the crop diversification. But the coefficient of fertilizer consumption and GCA per RCMS are not statistically significant. On the other hand rainfall has a negative impact on crop diversification and its coefficient is statistically significant.

In 2002-03, factors considered for explaining crop diversification could explain only 53.29 per cent of the variation in crop diversification index. It is observed that the factors that were exercising positive influence on crop diversification continues to do the same in the year 2003-04.While the coefficient of fertilizer is found to be statistically not significant, that of the other factors are statistically significant .The only factor retained in the model having negative impact on crop diversification is rainfall and its coefficient is not statistically significant.

CONCLUSION

The main findings that emerged from the above discussion are as follows:

There is crop concentration rather than crop diversification in the state. The pattern of cropping has shifted in favour of cereals during the study period. Even after more than one and a half decades of economic reforms, commercial crops have not made any dent in the state except in a few districts.

The factors that have significant influence on crop diversification such as fertilizer consumption, Proportion of irrigated area, GCA per tractor/ trailor, crop loans and GCA per RCMS are found to be weak in the state for which there is crop concentration.

Specific suggestions that we want to make in the interest of efficient management of soil and water resources particularly in a poor state like Orissa are:

(i) Government should take immediate steps to encourage crop diversification in the state.

(ii) NGOs and Panchayati Raj Institutions may be involved in creating awareness among the farmers regarding the importance of management of natural resources in general and soil and water management in particular.

(iii) Funds for Research and Development be earmarked so as to identify crops best fit for soil and water in different regions.

(iv) Finally, Foreign Direct Investment may be encouraged for development of agricultural infrastructure so as to facilitate agricultural diversification on the one hand and soil and water resource management on the other.

REFERENCES

Sekhar C.S.C. (2003),"Agricultural Trade liberalization–Likely Implication for Rice Sector in India", *Indian Journal of Agricultural Economics (IJAE)*, Vol. 58, No. 1, Jan-March.

Saleth, R. Maria (1997), "Diversification Strategy for Small Farmers and Landless:Some Evidence from TamilNadu", *IJAE*, Vol. 52, No. 1, Jan-March.

Sharma H.R. (2005), "Agricultural Development and Crop Diversification in Himachal Pradesh Understanding Patterns, Process, Determinants and Lessons", *IJAE*,Vol. 60, No. 1, Jan-March.

Gopalppa, D.V. (1996), "Crop Diversification and Income levels in Karimnagar Districts of Andhrapradesh", *IJAE*, Vol. 51, No. 3, July-Sept. 1996.

Vyas, V.S. (2000), *Changing Contours of Indian Agriculture*.

Dr.Utpal Kumar (2003), "Changing Cropping System in theory and Practice: An Economic Insight into the Agrarian West Bengal", *IAJE*, Vol. 58, No. 1, Jan-March.

Chadha, G.K. (2003), "Indian Agriculture in the New Millenium: Human Resources to Technology Challenges", *IJAE*, Vol. 58, No. 1, Jan-March.

Kalamkar, S.S. & Pradhan, P.C. (2003), "Diversification of Agriculture in Orissa: Progress and Possibilities", *Agrarian Transformation in Orissa: Issues and Evidence*, Vol. I.

Tripathy, P.K. Bhoi, T. Panda Pravati (2004), "Agricultural Sector in Orissa: Growth Profile and Crop Pattern Changes", *Agrarian Transfornmation in Orissa: Issues and evidence*, Vol. II.

Ratnam, C. & Mohan, Geetha (2005), "Crop Diversification in Andhra Pradesh: Trends and Determinants", *Souvenir*, 88th Annual Conference of the Indian Economic Association Dec. 2005.

Jalal, R.S. (2005), *Agreement under WTO and Need for Agricultural Reforms*.

Dasgupta, P.S. and G.M. Heal (1979), *Economic Theory and Exhaustible Resources*, Cambridge University Press.

41

INTERLINKING OF RIVERS IN INDIA : SOME REFLECTIONS

B. Eswar Rao Patnaik and Santosh Patnaik

SECTION - I

INTRODUCTION

It is possible to state that, river linking project is not altogether novel to water resources personnel and scientific community of the country. The eminent pilot Dastur has added fresh materials to the body of linking river projects by developing the garland canal concept. Arthur Cotton's insistence on linking rivers for accelerating the pace of transport development of the country deserves mention. As far as in 1960 Dr. K.L. Rao has published his thought provoking work, "Garlands of Rivers Project", which has mooted on linking the rivers of the country. Pioneering intellectual work of noted thinker, Annand Swaroop Arya has enriched the literature on linking rivers.

The Process of linking rivers Raine-Rone, Raine Danube has enabled the European nations to climb high on socio- economic ladder. The secret of the story of amazing agrarian and transport development of Chinese economy is linking rivers project. It is admitted on all hands that, the Indians backed the perception that conceives rivers as national assets, irrespective of the regions in which they flow. The sharing of waters of rivers Krisna and Yamuna has been the bone of contention between States, Andhra Pradesh and Karnatka. Global action may be guided on conservation principle which Stresses that water may be abundant in one place but it is scarce in others. Water is a fundamental right and this yellow gold of the 21st century must be provided equitably to all people and not on the basis of ability to pay.

Commenting on the N.D.A Government's decision on interlinking rivers, Dr. N.K. Rath (1993) has observed that, very little information is available in published form to interested students about project linking of rivers in India. Analysts argue that the programme for linking of rivers does not form an integral part of the Tenth five year plan in India.

A.Vaidyanathan (1993) questions the necessity, desirability and feasibility of linking of peninsular rivers, which involves an outlay of Rs. 5,60,000 crore. The answer to the question how the government can mobilise colossal resources to implement the project escapes our logic. Recently, Suresh Prabhu, the chairman of the Task Force for implementing river projects has candidly admitted that, even preliminary feasible studies have not yet been made. So, it is imperative that, an in depth study of environmental impact, displacement of people and submergence of villages consequent upon linking rivers is made.

SECTION - II

GANGA–CAUVERY LINK

The principal objective underlying Rao's work "garland of rivers projects" was to establish a link between rivers, Ganga and Cauvery. A sensitive approach to.water concerns in designing and implementing water development programmes is a must for reserving water scarcity in regions like, Telengana and Rayalaseema districts of Andhra Pradesh. Rao has focused on fourteen major rivers of the country namely Sindh, Ganga, Brahmaputra, Sabarmati, Narmada, Tapi, Subarnarekha, Brahmani, Mahi, Mahanadi, Godavari, Krisna and Cauvery.

Rivers have been classified into three categories by Rao. The first category rivers included ayacut covering a distance of 20,000 sq.km. and above. Ayacut covering a distance of 2,000 sq.km. to 20,000 sq.k.m. included second category rivers. The third category rivers included/ayacut covering a distance of less than 2,000 k.m. Central to Rao's proposal is the construction of a canal that links rivers Ganga and Cauvery covering a distance of 2400 k.m.

The link was to take off near Patna, Pass through the basins of rivers Sone, Tapi, Narmada Godavari, Ponna and join the Cauvery upstream of the anicut. The general impression was that the sink would withdraw 60,000 cusecs of water from the flood flows of Ganga for about 150 days in a year and presumably it may entail a lift of a sizeable part of the water above 450 meters. It appears that, the project may result in transfer of water from Brahmaputra and Ganga westwards to Southerhn U.P., Harayana, Rajasthan and probably southwards to the peninsular components. The peninsular components embrace several links, of which the premier ones would be those connecting the Mahanadi, the Godavari, the Krishna and the Cauvery.

The National water Development Board which was directed to enquire into the feasibility of the innovative proposal of Rao has recommended steps for linking rivers above Vindhyamountains in Southern region. The initiative of linking rivers below Vindhyamountains was not advocated by the Water Board, owing to its financial implications.

SECTION - III

BENEFIT- COST PRINCPLE AND CASE FOR LINKING RIVERS

The Social benefits of a Project undertaken includes, pecuniary and real benefits. Real benefits accrued to the consumers by a Project are reflected in addition to community welfare. Pecuniary benefits denote increase in earnings consequent upon changes in relative prices of factors and products (see table 2). While tangible benefits are evaluated in terms of market prices, intangible benefits and costs are evaluated in the form of effects on the health and living standard of people, using shadow prices.

Analysts agree that, the ratio of net present value of the project (NPVP) or the benefits cost ratio provides a measure of economic feasibility of the project. In case of a single project, it is choosen in its B/C ratio is one or grater then one. In case of several competing projects, ranking of projects are in descending order, on basis of B/C or B-C/C. However, future uncertainties need considerations, while using internal rate of return, so that, present value = future value X discounting factors. The social marginal productivity of a project is evaluated by deducting social opportunity cost of displaced persons from marginal product.

SMP= MP-SOCIAL OPPOURTUNITY COST

Case for River linking Projects

It is imperative that, dams and reservoirs are built in the economy to accelerate the process of inter linking rivers.

The insistence of planners and policy makers on dams derives, its strength from the fact that, agriculture till today, provides sustenance to 60% of people of the country. The lacklustre performance of plan exercises in irrigation frontiers is evident from the fact that, as high as 60% total cultivated land does not have irrigation facilities. So, since independence, the extension of irrigation facilities in farmer's field in rural and urban areas has been an important component of national endeavor. One bright spot in the body of river linking project is the possible improvement in the per acre yield of crops and agricultural progress, which may attend the creation and utilization of irrigation in an economy.

Data analysis of the estimate of central water Commission (Table 1) places the gross volume of water which includes both ground water and surface water transferred out of five basins, Mahanadi, Godavari, krisna, Perriyar and Cauvery at 65 billion cubic meters. The figure works out to 62 bcm. Providing for enroute losses to the extent of 2.7 bcm. The gross value realized in the transferee basin is 39 bcm., much of which is transferred from one basin to another. Allowing for this, net availability in the five basins increase by about 24 bcm. It is clear, therefore that, the projected expansion of irrigated area at the national level may be related to the increasing requirement of food and fibre production for a growing population.

Power Generation

Power sector development in a country is the hall mark of modern civilization. As an economy develops, there will be enormous demand for use of electricity, from all quarters, namely, agriculture, industry and transport sectors. So, expansion of capacity generation is indispensable to cope with the surge in demand for the use of power. One feather in the cap of mammoth multipurpose projects is the creation of capacity in power generation. A case of study of power sector development in Orissa reveals that, the five Hydro-Power projects in state sector, Hirakudh, Balimela, Machkund, Rengali and Upperkolab will create an installed capacity of 37.5 megawatts, 360 M.W. 34.5 M.W., 250.0 M.W and 320.0 M.W, respectively.

Table 1
Increase in Availability and its Disposition Arising from the Interlinking of Pennisular Rivers as Proposed by Central Water Commission (bcm)

Basin	*Valume transferred out of basin*	*Enroute utilization*	*Losses*	*Volume transferred out of basin*	*Net Increased in availability*
Mahanadi	11.2				-11.2
Enroute		-3.9	.80		
Godavari	25.5			6.5	19.
Enroute		7.5	.68		
Krishna	16.5			17.3	+08
Enroute		5	.75	(-)	
Ponmar	8.6			11.5	+2.9
Enroute		3.2	.39		
Caouvery	2.3			5.0	+27
Below Cauvery		2	1		
Total	64.1	21.6	2.86	42.3	+6.4

Table 2
Illustration of Project and costs in Evaluation of Efficiency Criteria of Public Expenditure to an Irrigation Project

		Benefits	*Costs*
Real			
Direct	Tangible	Increased farm Output.	Cost of Pipes
	Intangible	Beautification of Area	Loss of Wilderness
Indirect	Tangible	Reduced Soil erosion	Diversion of water
	Intangible	Preservation of rural Society	Destruction of wild life
Pecuniary	Relative improvement in position of farm equipment industry		

Employement Generation

It is plausible to believe that, planning is an instrument for betterment of people's living conditions. The generation of large scale employment opportunities in an economy will fight the battle of abysmal poverty.

The line of reasoning of advocates of plans is that, the field for employment base of the economy stands to the broadened, thanks to the policy thrust on dam construction. The objective is realized by construction of roads, creation of transport links and growth of economic activities, like opening of sales and purchase outlets, pisciculture and set up of secondary and ancillary industries, which is a logical corollary of construction of dams.

ABATEMENT/MITIGATION OF DROUGHTS

Human memory is alive to the nightmarish droughts of 1987-88 of country level, which has acted as barrier in the path of agricultural progress of the economy and inflicted untold miseries on the people of the country. It is well recognized that, severe droughts are recurring phenomenon in Kalahandi District of Orissa and Rayala seema District of Andhra Pradesh. Even now, there has not been any progress in the direction of sagacious and meaningful drought management in the country. It has been well reasoned by protagonists of mega projects that, the creation of mini watersheds and water-harvesting structures in an economy may be an answer to the challenges of droughts. The provisions of water to water scarce regions may facilitate the pursuit of rabi crop or kharrif crop in an area and to that extent, it may lighten the burden of drought on the economy. Micro-watershed management involves total resources conservation including soil and water, forest, horticulture and crop husbandry.

The case for dams rests on safety, they provide against floods. This is achieved partly by storing surplus water of rivers and partly by diverting water to water- stressed regions of the country. History bears testimony to the fact that , the Cuttack District of Orissa was saved from the wrath of floods, times without number by Hirakud dam.

CASE AGAINST DAMS

Higher Costs of the Projects

The official estimate does not specify the rate of interest of the inflation in estimating the total cost. Dr. N.K. Rath (2003) assumed a 7% rate of interest to be charged on the capital during the construction period. The total cost of the Himalayan, the Hydro electric and peninsular will come to Rs. 11,478.73 instead of Rs. 5, 60,000/- crore as originally estimated.

Unworkable Nature of the Scheme

A. Vaidyanathan (2003) reasons that, the project may comprise 35/30 odd projects, each of which, may have to undergo estimation and clearance. We may not be sure in advance that, all of them will be cleared and if so, it is unclear how the total mega project can be made.

The river-linking project is not the answer to the major needs of arid zones and drought prone areas of the country. The line of reasoning presented by A. Vaidyanathan runs thus:

The project does not purpose to take water from rivers to river less areas but merely to Link Rivers. By linking river A to river B by gravity we can succeed only in transferring some water to river B, which will then flow down hills again by gravity. This may make a little water available to areas already served but how this takes water to India's uplands and dry lands needs analysis.

Submergence of Villages

In the present century, environmental degradation has emerged as a global concern for human survival. The manmath irrigation project will submerge villages, after installation and tribal population living in village areas would lose their land. Fernandos (1994) has estimated that, during 1951-90, 213 lakh people were displaced because of development projects. The sad tune in the song of Narmada project is submergence of 37 villages comprising 1000 families and 3.5 lakh hectares of India's best deciduous forest comprising teak and bamboo plants will be submerged in the process. It seems that, the cost benefit analysis of projects made by experts, takes into account only tangible economic values, like timber and minor forest produce and discounts invaluable ecological values, like water replenishment, wild life shelter, siltation and climatic stabilization. Reference may be made to the recent promise made by Chief Minister of Orissa, that "Records of Rights" of oustees of Huirakudh Dam would be provided soon. The policy of government should be as for as possible to give land in exchange of land and that too well in advance of submergence.

It would not be out of contex to mention here that the task force for implementing Linking Rivers project does not have an independent professional secretariat. The organization is manned by professionals from irrigation bureaucracy who have not shown much interest in developmental works.

It is an established fact the river valley projects are not a positive proposition in India. Usually a project is sanctioned by a planning commission only, when the cost benefit ratio is 1:1.5. Numerous studies on country level have shown that, the cost benefit ratio is 1:0.46. The suggestion, therefore, is that big dams are not, as beneficial as claimed by proponents of big dams.

CONCLUSION

The foregoing analysis has presented the balance sheet of river valley projects. Major dams have brighter as well darker sides. In lively debates questions have been raised by scholars of all hue regarding the viability of high dams. The alternative suggested is construction of a number of smaller dams instead of one dam or smaller structure. If mega projects become millions of small projects full economic costs are likely to be recovered.

Minor irrigation works and small dams with minimal impacts of water logging and salinity deserves attention. The success that dams can achieve in mitigating the wrath of floods, largely depends on careful maintenance of embankments for avoiding breaches in embankments, and preparation of inundation maps for every breach of embankment. The perception that rivers belong to the nation has to be understood by the people of the country. The daunting task of public authorities is to provide for an effective rehabilitation policy for project affected persons supply land for land, ensure employment of the oustees in suitable lines, fix compesation for loss of land at just price and accelerate the development of social and economic infrastructure around the places, where oustees are rehabilitated. All this depends on the collective wisdom of policy makers public authorities and irrigation experts. The various aspects of the project for interlinking rivers, conceptual and technical, economic and environmental need judicious and meticulous review by independent experts. The Bruntland Commission counsels us "The world in which poverty is endemic will always be prone to an ecological and other catastrophes". As the Vedic seers state succinctly Aapsu amrutam".

Of late, the accent of the economy has been on globalization and liberalization, which have by-pass social welfare considerations for the sake of profit maximization. It seems that, the proposals of major irrigation projects are underway in the economy. Analysts, policy makers and socially oriented persons of the country should raise the scale of endeavor to consider the balance sheet of mammoth irrigation projects and evolve alternatives to costly and environmentally less friendly dams.

BIBLIOGRAPHY

A.Vaidhyanathan (2003), Interlinking Pennisular Rivers: Acritique. E.P.W. July. 5.

Ramaswamy Aiyar (2003), Letters to Editor E.P.W. July 1.

N.K. Rath (2003), "Linking Rivers": Some Eleminatray Airthmatic E.P.W. July.

E.E. Nadu (2002), "Nedu Jala Sandhanam Sudura Swapnana", Himabindu 10 August.

Eswar Rao Patnaik (2003), "Large Dams, Sustainable Development and Problems of Displacement". S.N. Tripathy (ed.) Economic Development and Problems of Displacement, 2003, Anmual, Publishers.

Editor, Indian Express, July 18, 2004.

Interview with B.Krishna Rao Patnaik, "Executuive Engineer" Khariar.

S. Anand Krishnan (2003), "Is the water Bomb Tickling for India". India Today May.

S.N. Behera, "Displacement and Development. A Cost Benefit Analysis, Published in Economic Development and Problems of Displacement (ed.) S.N. Tripathy.

42

UTILIZATION OF LOCAL RESOURCES FOR SUSTAINABLE HORTICULTURE AND VEGETABLE CULTIVATION

B.M. Kumara and S. Rajendran

In India the Agriculture system has been undergoing tremendous changes starting from subsistence farming to post revolution over the years. Though the share of this crucial sector to the total economy has been on the continuos decline, still a large chunk of population either directly or indirectly involved in agriculture for their very survival. In the process there has been a sea change not only in the cropping patterns but also in the farming practices. The green revolution packages paid more attention towards the cereals and due to various factors other crops including horticulture and vegetable received enough concentration very much later. Alongside the green revolution, due attention was not given to utilize the inputs like land, water and chemicals more judiciously and as a consequence, environmental problems such as water logging, salinity and desertification were observed. In fact this has results in reducing the yields in some crops in assured irrigation regions. To overcome this, the concept 'sustainable farming system' has been advocated not only in India but in other regions in the world as well.

According to the Technical Advisory Committee of the Consultative Group on International Agricultural Research (1998) *"Sustainable agriculture is the successful management of resource's for agriculture to satisfy changing human needs while maintaining or enhancing the quality of the environment and conserving natural resources"*. However, concrete methodology is yet to be developed to measure the agriculture sustainability and hence it is a Herculean task to comprehend this complex concept for analytical exercise. Against this, in the present paper an attempt has been made to understand the utilization of locally available natural resources and the ecological and environmental implication on the horticultural including vegetables among the 17 farms, which practise sustainable farming principals all over the country (For a methodology see Rajendran, (1998). A word of caution is essential here as the farms are located in far off places with extremely diversified agro-environmental condition it is very difficult to generalize the observations, albeit this given an opportunity to understand the complex issue with field based data. The objectives of the paper are twofold; firstly, it makes an attempt to study the pattern of resource use and farming techniques among the farms and secondly, it examines the economic and environmental impact on the farms.

All the farms have ideologically committed to utilize the locally available farm grown inputs for growing different horticulture and vegetable crops besides growing other crops. In other words all the farmers have decided to use the local–crop–residues and other farm based products as inputs for growing horticulture and vegetable crops to reduce the input costs and maintain clean environment. Adoption of diversified resource base including animal husbandry makes a farm to be self-reliant and Table –illustrates the resources generated in the farm itself.

Table 1
Particulars of resource Generation in the Farms

Base for Resources	*Resources Generated*
1. Treeslegumes, creepers and medical plants	Green manure, mulching, wind cover, biopesticides, fuel fodder and timber
2. Livestockbovines, ovines and poultry	Munure, cowdung cake, biogas, slurry and cattle urine
3. CropsHorticultutr, Vegetable, cereals and others	Crops residues, fuel, compost, mulching, food, fruits, vegetables and others
4. Natural predatorsowls, vultures, snakes and spiders	Check pests and diseases

The table one very clearly demonstrates how a resource rich farm can generate different resources which can be used more on sustainable basis. This not only brings down the dependency of the farmers for external inputs also the inputs costs. All these help the farms to successfully undertake the sustainable farming system on a sustained basis for maintaining food security and self–reliance.

In addition to this, there is a vast potential to export the horticultural products, produced through eco-friendly techniques particularly after the Indian agriculture has been opened up for global market. Available statistics indicate that of the agricultural products exported fruits and vegetables accounted for 2.1 per cent during 1960-61 and this has increased to 4 per cent during 1992-93 which show an impressive 3.2 per cent of compound growth rate. In fact due to health consideration and other environmental reasons, the consumers especially in the west are ready to pay premium price for the horticultural products which are being produced strictly with natural inputs, avoiding synthetic chemicals. In this context, India has a comparative advantage in the global market as this country can grow variety of horticultural products due to itd diversified afro-environmental factors.

Unfortunately, neither the policy makers nor the farmers pay attention to this and hence, the application of synthetic is on the increase, which has been observed over the years. For instance, as per the statistics from the fertilizer association of India the use of fertilizers has increased manifold merely from 1.93 kg/ha during 1960-61 to 77 kg/ha during 1996-97. though this quantum is not too alarming as compared to world levels, the farmers use the fertilizers in an indiscriminate manner which causes more harm than benefit. Simultaneously application of plant growth hormones, fungicides and weedicides is also showing an upward movement, which again harms not only the environment but also the health. Obviously, all these lead the rich consumers to opt for naturally grown, chemical free horticultural products and there are agencies at international level to rigorously monitor, inspect and certify the products. In the following the farming practices under sustainable system are discussed based on the field experiences.

FARMING PRACTICES

Farm practices under sustainable farming are completely different from the modern or conventional farming. Mixed or intercultivation is practiced under sustainable farming to avoid total crop failure and loss. Local dry leaves, grass, farm wastes, cow dung and green manure are used on maximum quantity to enrich the soil fertility and thereby increase the production of crops, which makes the soil more porous and the beneficial organisms like, earth worms proliferate naturally. All these make the soil fertile which is sustainable and the yield levels are sustained in the long run. No local byproducts are allowed to go waste or burnt. In other words the farmers mere attention to recycle the farm wastes to sustain the soil fertility. And sometimes, silt and coir dust are brought from outside the farm to apply for increasing the soil fertility. A couple of farmers prepared their compost within the farm itself NADEP or Indore method. According to these farmers unlike the application of factory made synthetic inputs, such practice increase the employment opportunities of the farm workers.

Such methods as per the farmers, have not only increased the soil fertility, water holding capacity but also help maintain sustained yield levels. During the survey some farmers reported that a meager amount of 20 liters of water were used per plant (sapota and coconut) under trench method (popularly used under sustainable farming), 70 liters were used in drip irrigation technique and 150 liters were used under flood system. As is a known fact that a maximum of Rs 50,000 is required for installing drip irrigation system on one ha plot. This clearly explains the magnitude of conserving and economizing the scarce resource–water–in water scarce and river water dispute driven country like India.

In addition to these, all the surveyed farmers, use plant based biopesticides to combat pest and diseases in the crops. Though this appears to be simple, in practice, this is a highly complex and experience oriented activity, it requires patience, commitment and careful selection of plans and their proportions. All these are meticulously done by the members particularly the younger ones of the farmers. The plant besed repellents are prepared from the popular plants such as *errukkan (Calotropis gigantica),* mari gold. *Pongamia (pongamia globra), tusi (Ocimum sanctum)* and neem *(Azadirachta indaca).* Leaves of these plants are normally choked in the water mixed with cattle urine sometime diluted with neem oil for a period of one week either in the pots and plastic containers or cement tanks to get strong odor. Depending upon the availability of these plants in a given season the farmers venture into preparing the decoction. Inspite of the fact that all the farmers preferred to spray the decoction through sprayers or long plant leaves on the plants, it was practically difficult to apply on the well grown–taller–pomegranate, mango and ber trees. Farmers readily keep the decoction with them for application time and again on the plants sometimes merely in anticipation of pest attack. Since there is no scientific formula and prescription unlike available for synthetic pesticides, the excess application of such decoction may lead to harmful effect on the plants and fruits. A farmer near Banglore admitted and suspected that excessive spraying of biopesticides resulted in block dots on the guava fruits. Nevertheless, the farmers confessed that ecofriendly biopesticides not only controlled the pests like caterpillars and diseases, brought down the external input costs but also the self life of the fruits and vegetables increased considerably. The interaction with different agricultural scientists from Solapur and Coimbatore has also confirmed this fact.

Intercropping or mixed cropping of horticultural crops is also adopted on a large scale under sustainable farming to avoid crop loss. This has been observed more among the vegetable cultivation where more than half a dozen crops are grown at a time on a single plot. For instance, tapioca, turmeric, ginger, onion, tomato and a host of other vegetables were grown on different rows which have made the farmers to get assured source of income. The farmers found that mixed cropping not only ensured income from the successful crops but also particular crop serves as repellent to other crops of the same plot.

Another sustainable farming practice for the horticultural crops is pruning. Through proper planning for pruninh the plants, one can change the fruit yielding patterns of the crop. It is a well known fact that the farmers get very low share from the consumers' rupee and this is very high during the peak harvest season of a particular crop. As a result neither the producer nor the consumer benefits and the middlemen who grab a lion's share. In such a circumstances, pruning helps at least some crops like ber to adjust the flowering the ultimately fetch more income. Pruning of pest infected and disease affected leaves, twigs and flower parts also prevents further spreading of pests and diseases and ultimately increased the crop yield and income. A study conducted among equal number of seven from ecological and reference (modern) farms in south India clearly revealed that (Table 2) on many counts the ecological farms performed better that their counterparts.

Table 2
Performance of Selected Variables Across Ecological and Reference Farms

Variables	*Ecological farms*	*Reference Farms*
Gross income (Rs/Ha)	20,430	18,340
Variable cost (Rs/Ha)	9,810	6,830
Gross margin (Rs/Ha)	10,620	11,515
Net farm income per labour day (Rs)	32	32
Labour days/ha	346	333
% of total produce sold	57	52
Case part of total costs (%)	49	67
Off-farm income (per farm/Ha)	15,660	6,950

Source: Jager and Werf, 1992, p. 47.

The table one clearly collaborates our finding. Very importantly, though the variable cost is found to be high among ecological farms, their case part of total costs is less than 50 per cent. Secondly, it is striking that the off-farm income is almost three times higher among ecological farms than then the reference farms, which clearly indicates that the ecological farms can provide sustainable income and employment to the farm sector. Though the sample size of the study is very small, it gives some clue to the economic viability and employment sustainability of ecologically sustainable farms.

CONCLUSION AND POLICY SUGGESTIONS

Broad issues on the patterns of local resource use among the farmers who practise sustainable farming have been discussed and based on these it is quite evidence that there is a potential to at least take up this on experimental basis across different agroenvironmental zones. The farmers are using the locally available resources more on trail and error basis and training by some NGOs really helped them to a larger extent. Similar experiments have to be taken up by the state agencies and such issues can be disseminated to the interested farmers. In some places, ecofriendly inputs like vermicastings are adultrated and sold to the farmers at high costs and strict market regulation will prevent this racket. In fact, very surprisingly, the supply of ecofriendly inputs like biofertilizers is very much short of demand as reported by Rajendran (1995).

Since there is a greater potential to expand the export market for horticulture products and vegetables, produced through sustainable farming, attention must be paid to generate foreign earnings besides fulfilling the growing domestic market. In fact it has been reported in the media that outlets, selling, organically grown horticultural products are on the increase and due to the absence of proper regulations, the customers need for regulation such outlets to safeguard the consumers. Even for export of such commodities severe inspection and certification is essential to establish the market claim in the world.

REFERENCES

Janger a de and Werf van der E. (1992), *Ecological Agriculture in South India, Agriculture Economics Research Institute,* The Hague, The Netherlands.

Rajendran S (1998), *An exploratory Study on Organic Farming in India, in Living Resources for the Millennium 2002,* ed. S.J. William, Loyola College, Chennai.

——— (1995), *Ecofriendly Biofertilizer–Need of the Hour, the Hindu Businessline,* February 22, Madras.

43

ENERGY AND SUSTAINABLE AGRICULTURE DEVELOPMENT IN INDIA

A. Marimuthu

Sustainable development has become an integral part of the terminology of development jargon. The most profound definition of it is the development which meets the needs of the present without compromising the ability of the future generations to meet their own needs. In India, agriculture, at least for a few more decades has to play a major role in economic development of the country as the main source of employment and income besides meeting the increasing demand for food and raw material. Since, the scope for expanding area under cultivation is very much limited, output growth in agriculture depends on improvement in productivity.

ENERGY AND DEVELOPMENT

Energy is of diverse kinds; most of them are nature endowed Bible claims that energy in the form of light is God's gift. Energy is equal to mass, multiplies by the velocity of Light Square (Einstein). Energy is one of the crucial inputs for enhancing productivity. The rapid increase in agricultural production in some of the developing countries including India in recent decades has resulted from increased use of commercial energy in the form of fertilizers, oil and electricity. However, indiscriminate and inefficient use of energy leads to environmental degradation, which is a major threat to sustainable agriculture.

In the classical economic theories of growth, energy was not even a distinct input in production. Today, energy is one of the factors limiting world development, although it is essential for the growth of production and productivity.

To bring about increase in agricultural production and also to increase employment opportunities in agriculture, the Five Year Plans in India brought additional land under cultivation, expanded irrigation facilities and used increasingly chemical fertilizers and pesticides and high yielding hybrid seeds etc. all collectively known as the New Agricultural Strategy.

Development of agriculture and industry has been accompanied by development and expansion of infrastructure–namely energy, transport and communication, banking and finance.

Agricultural producers add energy to the natural biological production process to rise the productivity and to reduce the variability and risk in the biological production process. In the primitive agricultural system, the process of energy transformation was carried out virtually in their normal state, i.e., the principle energy source was solar which was augmented slightly by human and animal inputs. But, in modern agriculture system to achieve greater levels of output, other energy inputs are used to enhance solar energy transformation. Table 1 depicts the infrastructural development in India.

Table 1
Trends in the Performance of Infrastructure Sectors in India

Particulars	*Unit*	*2003-2004**	*1950-51*
1. Coal	mt	389	32
2. Electricity generated	b. kwh	558	5
3. Petroleum–Oil crude	mt	33.4	0.4
4. Steel	mt	36.9	1.0
5. Cement	mt	123.4	2.7

* Figures are provisional
Source: Economic Survey (2004-2005) Ch. 9.
Data: Statistical Outline of India, 2003-2004.

From the Table 1, the coal production rose from 32 million tonnes to 389 million tonnes between 1951 and 2004 and power generation and non-conventional power plants rose from 5 billion kwh to 558 billion kwh and production of petroleum crude from 0.4 million tonnes to 33.4 million tonnes.

ENERGY, ENVIRONMENT AND SUSTAINABLE DEVELOPMENT–CERTAIN INDICATORS

Energy Economists opined that energy consumption is an index of economic development. In overall economic development of a nation, energy is essential to ensure adequate and sustained supply for every economic sectors of the economy. Efficient infrastructure is a pre-requisite for mobilizing economic development. Modern economists such as Schurr, Baly and Jorgen stated that energy is the fifth factor of production in addition to land, labour, capital and organization.

The per capita consumption of energy for selected countries is presented in Table 2.

Table 2 clearly explains that the first three countries are developed countries with high per capita incomes, while the next three are developing countries with low per capita incomes. Per capita consumption of energy in India is only 12 per cent of that in Japan and only 6 per cent of that in

U.S.A. Although commercial energy consumption in India has been steadily going up during the last decades, it is still much lower than that of developed and even developing countries.

Table 2
Per Capita Consumption of Energy for Selected Countries

Country	*Per Capita Consumption of energy (Kgs. of oil equivalent) 2001*
U.S.A.	7996
Japan	4099
U.K.	3982
Egypt	737
Indonesia	729
India	515

Source: World Development Report, 2005 and World Development Indicators, 2004.

There is deep correlation between energy use on the one hand and development, national income, standard of living and quality of life on the other hand.

Table 3
Energy, Environment and Sustainable Development – Indicators

Country	*Population 2001 (in million)*	*HDI 2000*	*Per Capita 2000*		
			GNP (in $)	*Energy Consumption (Kgoe)*	*CO_2 Emissions*
America	285.3	0.939	34142	8148	19.7
Bangladesh	133.3	0.478	1602	142	0.1
Britain	58.8	0.928	23509	3962	9.2
Canada	31.1	0.940	27840	8156	14.4
China	1271.8	0.726	3976	905	2.3
Ethiopia	63.8	0.327	668	291	0.1
India	1032.4	0.577	2358	706	1.1
Japan	127.0	0.933	26755	4136	9.1
Malaysia	23.8	0.782	9068	2126	5.4
Nepal	23.6	0.490	1327	343	0.1
Pakistan	141.5	0.499	1928	463	0.7
Singapore	4.1	0.885	23356	6120	13.7
Sri Lanka	18.7	0.741	3530	437	0.5
Switzerland	7.2	0.628	28769	3704	5.7

Sources: World Bank: World Development Indicators, 2003,
World Development Report, 2005 and
VNDP: Human Development Report, 2002.

It is inferred from the Table 3 that those countries with higher per capita energy consumption emit a large quantity of CO_2. The American population just 5.0 per cent of the world population consumes 28.9 per cent of the world primary energy consumption.

It is also inferred from the Table 3 that those countries with higher per capita GNP and energy consumption are also the countries that the technologically inclined. The less technologically oriented countries are larger to improve their standard of living and many recognize energy as being a necessary ingredient. This is the motivation, for instance, for the development of Non-Conventional Energy Sources in India.

POLICIES FOR SUSTAINABLE AGRICULTURE

(a) Use of Land

- Certain type of land use may appear to be highly profitable in the short-run but leads to serious land degradation in the long-run. Cost of such consequences would outweigh the short-run profitability.
- Effective implementation of laws prohibiting diversion of fertile agricultural lands to non-agricultural purposes is necessary at least to maintain the present level of net sown area.
- Priority should be given for soil conservation and reclamation of degraded soils.
- Introduction of appropriate insurance policy to minimize the effect of risk and uncertainty on soil conservation structures and crop production will provide conducive environment for more investment on soil conservation.
- Generation of location specific least cost technologies and policies to relax the socio-economic constraints.
- By providing adequate institutional credit, farmers should be educated about the long-run benefits of soil conservation.
- Organisation of farmers at village level could help in successful implementation of soil conservation and waste land development programmes.

(b) Use of Water

- Suitable electricity pricing policy is required to save ground water and electricity constraints in adoption of drip irrigation and other water saving technologies have to be removed by suitable policy measures.
- Soil conservation and construction of percolation ponds will enhance ground water recharge.
- The existing policy of subsiding canal water and electricity for irrigation does not encourage factories to conserve water and improving the efficiency of water use in agriculture.

(c) Use of Energy

- Energy is a strategic commodity because it is essential for the economy of the country and the supply is uncertain or may present a risk. India is a net importer of crude oil and petroleum products, giving rise to energy security concerns and vulnerability to external prices and supply fluctuations. Imports of oil and coal have been increasing. The dependence of energy imports is projected to increase in the future. Estimates indicate that oil products will need 75 per cent of total oil consumption.

- Various studies on energy use have indicated the scope for energy saving in agriculture without significant reduction in net return. But, current input prices favour the use of commercial energy inputs.
- It is high time that India adopted a responsible national policy of natural resources and biodiversity conservation which is in tandem with its economic and social development policies (Singh, 1998).
- The use of fertilizers should be integrated through more economic, renewable and environment friendly organic manures, biofertilisers crop residues and given manure to improve the soil health and productivity. It is not conceivable that agriculture can deliver the expected outputs without modern technologies such as genetic engineering and biotechnology (Leissinger, 1999).

An Evaluation

- Sustainable agriculture can be defined as one that equitably balances concerns of environmental soundness, economic viability and social justice (Allen *et al.*, 1991). With the help of these, India's agriculture is meeting the increasing demand for food in coming decades. Sustaining productivity growth in agriculture is the only option available to meet the challenge.
- India possesses the largest livestock's population (445 million in 2003) in the world and second largest populous country (1030 million in 2002) in the world. The waste created by the livestock and human population can be converted into wealth in the form of biogas.
- India is basically an agricultural economy which generates 450 to 500 million tonnes of biomass that can be used for production of power through biomass based power plants per annum. The estimated total power potential from biomass is 19500 MW.
- India is the fifth largest wind power producer in the world with wind power generation capacity of 4500 MW.
- Indian cities generate 30 million tonnes of solid waste and 4400 million cubic meters of liquid wastes every year.
- Energy scientists in India discussed for Bio-diesel. If India could bring the waste land for cultivation of bio fuel plants, our country could attain self reliant energy economy which will lead to increase the productivity of agriculture.

In order to keep pace with population growth and the imperative of upgrading nutrition, the world faces the challenge of doubling food production within the next decades. The world population is expected to grow to 7.7 billion in 2020 from 6 million in millennium year rising serious concerns about how food demand will be met in the coming decades. Sustaining productivity growth in agriculture is the only option available to meet the challenge and suitable policies for population control and poverty alleviation will go a long way in conserving natural resources, reducing environmental degradation and sustaining agricultural production.

REFERENCES

Doering, O.C. (1977), "An Energy based Analysis of Alternative Production Methods and Cropping Systems in the Corn belt", Report of the Agricultural Research Station, Purdue University, West Lafyette.

Brundtland Report (1987), "Our Common Future", World Commission on Environment and Development, Oxford: Oxford University Press.

Sah, A.K. (1998), "Environmental Protection and Sustainable Agriculture", Employment News, Vol. XXIII, No. 22, pp. 1–2.

Singh Katar (1998), Sustainable Development: Some Reflections, Indian Journal of Agricultural Economics, Vol. 54, No. 1, pp. 6–43.

Pachauri, R.K. (1998), "Economics of Energy use in Agriculture", Indian Journal of Agricultural Economics, Vol. 53, No. 1, pp. 313–322.

Kombairaju, S. and Lokanadhan, K. (1999), "Energy, Environment and Sustainable Agriculture", paper presented in a National Seminar on Emerging Issues in Indian Agriculture, (ed.,) Subramanian G., *et al.*, School of Economics, Madurai Kamaraj University, Madurai, Tamil Nadu.

Leissinger M. Klaus (1999), "Biotechnology and Food Security", Current Science, 76 (4), pp. 488–500.

Monga, G.S. (ed.) (2001), "Environment and Development", Deep and Deep Publication Pvt. Ltd., New Delhi.

Ramasamy, S. (2006), "India Needs Energy Security and Energy Independence: Some Reflections", paper presented in a State Level Seminar at A.N.J.A. College, Sivakasi, South Tamil Nadu.

Ruddar Datt and Sundharam, K.P.M. (2006), "Indian Economy", S. Chand and Company Ltd., New Delhi, pp. 114–122.

44

GROWTH AND SUSTAINABILITY IN AGRICULTURE OF KARNATAKA

Channabasavanagouda P.

INTRODUCTION

Agriculture plays an important role in the economy of the state. Besides contributing a substantial part of the state Domestic product (about 28.20% in 1997-98), it is the largest source of employment (about 67%). Karnataka has an advantage due to the existence of ten agro-climetic zones, different soil types, bimodal rainfall pattern and irrigation potential. The state has potential for growing varieties of crops, increasing productivity, cropping intensity in rainfed/irrigated areas and also for increasing area under hybrid/HYV crops.[1]

The share of agriculture in gross state domestic product (GSDP) has decreased from 33 per cent to 25.3% in 2000-01. The growth rate of both net state domestic product (NSDP) and GSDP originating from agriculture has been much lower than the secondary and territary sector.

In spite of the lower rates of growth in agriculture, the state economy seems to be quite sensitive to the performance of the sector due to strong inter-linkages. The growth rate in agriculture, which was quite commendable till the end of 1970's decelerated during the 1980s. The production of foodgrains also stagnated. Productivity was low, both under rainfed as well as under irrigated agriculture, during the eighties. During the eighties and early nineties, with the help of state level policy interventions, the state achieved a recovery from the large coming stagnation.[2] A number of centrally sponsored thrust programmes were implemented like the intensive rice development programme, the pulses development programme and the oil seeds production programme focused on stepping up production and productivity on watershed basis. This paper tries to focus on growth of agriculture and prospects for sustainable development in Karnataka.

METHODOLOGY

The present study is based on the secondary data relating to agriculture production, productivity and area of Karnataka. The required data were collected from various issues of economic survey CMIE (Center for Monitoring Indian Economy) reports, for analysis. This study used timeseries data, percentage and tables.

AGRICULTURAL GROWTH

Land Using Pattern

Karnataka state with 19.10 million Ha accounts for 6.3% of India's geographical area. Area under forest in the state is around 16 % as against 22% at the national level but the share of net area shown is much higher in the state (55.62%). Than the national average of 47%. Expansion of cultivation on pasture and grazing lands, and wastelands have contributed to the increase in the net area sown by 0.53 million ha in the state between 1955-93.

It is not surprising that in the state, the percentage of area sown more than once (8.58% in 1995-98, up from 1.8 per cent in 1955-58) is lower than the national average (about 13%, up from 6.1 per cent) owing to lower level of irrigation development. More area under industrialization and urbanization has resulted in increasing the share of land put to non-agricultural uses almost by 50% both at state and national levels between 1955 & 1993.

Crop Shifts

The area under total cereals in the state stood at 45%, down by 11% in 1999-2000 as against 56% in 1955-58. The decline is mainly owing to reduction in area under Jowar and Bajra. The area under paddy increased by 3% (from 8.7 to 11.6%) between 1955-58 and 1990-93 and stabilized around that level after that. The area under maize increased from 0.1% in 1955-58 to 2.5% of the gross cropped area in 1990-93 and further to 4.31% in 1997-00. Though there is a decline in the proportion of area under cereals, paddy, jowar, maize and wheat, continue to be the dominant cereal and millet crops in the state.[3]

Area under pulses has remained more or less unchanged (around 14%) during the last four decades. However, the area under the oil seeds in the state, which was around 12% till the late 70s, increased to 24.4% after the 80s., but comes down to 19.6%. The area under groundnut and sunflower in the state shared about 4.6 and 1% of the total area under respective crop in the country. The area under other cash crops such as sugarcane, coconut, arecanut etc., has increased over the years.[4]

Production Trends

In the table 1 production trends of important crops have been analysed here with the help of compound rates of growth for production and productivity. It is observed that the production of cereals in the state has grown at 2.13% per annum during the period 1955-94. The entire growth has been contributed by the yield as area under cereals in the state decelerated during this period. The expansion in area under cereals during 1955-56 through 1967-68 resulted in significant growth in cereals production in the state.

Table 1
Compound Growth Rates in the Crop Economy

(% annum)

Crops	*1955-56 to 1993-94*			*1990-91 to 2000-01*		
	Area	*Production*	*Productivity*	*Area*	*Production*	*Productivity*
Rice	0.60	2.21	1.60	0.60	2.21	1.60
Wheat	-0.63	2.73	3.38	3.98	4.78	0.77
Jowar	-0.84	1.26	2.12	-1.28	-4.54	-3.3
Bajra	-0.47	2.19	2.67	-1.4	-9.18	-7.89
Ragi	0.44	2.7	1.72	-3.22	-4.32	-1.14
Maize	10.76	15.52	-	7.75	6.86	-0.83
Total cereals	-	2.13	2.43	0.14	-0.03	-0.17
Gram	1.02	1.80	0.77	1.02	1.80	0.77
Tur	1.42	1.94	0.52	1.5	-1.26	-2.72
Total pulses	0.83	1.79	0.95	3.92	-1.11	-4.84
Groundnut	0.65	1.55	0.90	-2.03	-10.4	-8.54
Total oil seeds	2.18	2.74	0.54	-3.81	-9.56	-5.98
Cotton	-1.60	2.39	4.06	-3.01	-3.42	-0.42
Sugarcane	4.40	4.87	0.45	-0.7	1.49	2.21

Source: Deshapande R.S. and J. Prachitha (2005), "WTO and Agricultural Policy in Karantaka.

The pulse production in the state accelerated by 1.79% per annum during the last 40 years and by 2.71% during the seventies. However, it grew at a rate of 7.33% per annum between 1980-81 and 1993-94. The performance indicators of the crop economy at the state level merge the differences across regions.

In the cropping pattern of the state, cereals and oil seeds are the two major crops followed by pulses. The principal cereal crop in terms of area is Jowar with 18.97 lakh ha followed by paddy with 13.53 lakh Ha (1997-98). In the total oil-seeds area of 23.72 lakh Ha, groundnut and sunflower account for 83% of the area. Pulses account for 16.82 lakh ha of area.

PROSPECTS FOR SUSTAINABLE DEVELOPMENT

There is a large un-exploited yield gap for many crops between what is achievable and what has been achieved. This has to be bridged on priority. Prevention of loses due to pests and diseases during production, inadequate storage and processing, grading, packing, commodity specific and region specific cost effective post harvest infrastructure and marketing system adds further to the kitty.[5] Other areas which after promise for further improving agriculture on a sustainable basis are outlined below.

Bio-technology

The further lies in bio-technology, be it is in the breeding and development of stress resistant field or horticultural crops or livestock production, theraputics, vaccine development prevention of crop losses etc. To achieve the goals of tolerance to biotic and abiotic stresses, improving the productivity with desired quality, biotechnological intervention through marker aided selection and gene transfer technology complementing conventional recombinant breeding in inevitable. But, it is cost intensive technology, which required adequate financial support from government and through private participation. The scientists concerned need wider exposure to basic sciences like molecular biology and bio-chemistry for harnessing the potential.[6]

Land Resource Management and Waste Land Development

Soil erosion, salinity, alkalinity and drainage impedance constitute the major causes of land degradation in the state. The extent of areas affected by different degradative process is increased, which need to be brought under use by suitable means. Technology are available, but concerted action among all stake holders is required.

Karnataka is one of the states having 4.3 lakh Ha (2.3%) waste land. Specific programmes are needed to harness this resource. Integrated watershed development programmes are key to bring the wastelands to productive use. People participation and their active collaboration are, however, essential to make the programme successful.

Irrigation and Groundwater Exploitation

Irrigation development in Karnataka as compared to many other states and the country as a whole is quite low. The gross irrigated area in Karnataka is 23 lakh ha. Until now, only 52% of the utlimate potential of all sources has been developed.

Information management, easy data base accessibility of soils, water resources, agrometeriology and water requirements of crops, integrated water management plans, conjunctive use of rainfall water, surface and ground water, education and awareness, ID crops to suit the water availability are the need of the hour.

Diversification of Agriculture

Diversification of agriculture is the key for dynamic, remunerative agriculture and employment generation. Horticulture related activities, dairying, poultry, sheep husbandary, piggery, apiculture, sericutlure, mashroom culture, fish culture, fodder production etc., along with agriculture should receive greater attention since agriculture alone will no more be remunerative.

IPM and INM

Increment use of fertilizer and pesticides could not be prevented or avoided despite the well known facts about their ill effects. There is need to minimize persticide usage by adopting Integrated Pest Management (IPM) and minimize fertilizer usage by going in for Integrated Nutrient Management (INM). Incidentally, both will reduce the cost of production of crops and drive agriculture towards sustainability. Research and development effort on the production and popularization of non-pesticide components of IPM such as botanical pesticides, parasites predators, bio-agents like viruses, bacteria, fungi etc., have to receive greater attention.[7]

Mechanized Agriculture

Complete mechanization of agricultural is neither feasible nor desirable, because of the instability of resource for poor farmers to afforded the costs of implements and machinery. Further, farm labourers also need employment. However, low cost, simple animal drawn and hand operated improved agricultural implements, which have been designed to reduce drudgery and increase efficiency of farm labour, have to be developed and popularized.

Agro-processing and Marketing

One of the primary aims of agricultural development is to enhance the farm productivity and ensure remunerative prices to the farmers. The whole gamut of issues related to marketing of agro-based raw material and finished goods need revamping. A large proportion of farmers would to be exploited by large number of unscrupulous middle men and market forces. Modernization of market yards, creation of storage facilities and access to reliable market information and market intelligence, harnessing information technology, go a long way in reducing the market problems of farmers.

Export Oriented Agriculture

Karnataka has vast scope for the export of cotton, sugar, tobacco and to some extent maize. Inspite of the availability of all the production potential, vast gap exist in yeild and technology transfer. This, together with the inadequate attention given for grading, packing, storage and processing, commodity specific and region specific cost effective post harvest infrastructure, unorganized marketing system etc., resulted in non-exploitation of maximum production and export potential. It is needed to improve the grading, packing, storage and processing technology and also needed to develop infrastructure and marketing system.

CONCLUSION

Agriculture of Karnataka has been characterized by intermittent phases of growth and stagnation. It is a matter of deep concern as this sector relates directly to the overall growth performance of the state economy.

Efforts, both at policy and at implementation levels, are being made to overcome the constraints and encourage growth boosters. The constraints mainly relate to efficient utilization of natural resources, existing infrastructure facilities, forward and backward linkages, and allied supportive activities. Besides these, current liberalization process has led to the emergence into prominence the role of allied sectors, e.g., horticulture, floriculture, fisheries, agro-processing, etc., these sector require better quality of supportive infrastructure for sustainable development and only then it can contribute significantly to the growth of Karnataka's economy.

NOTES

1. R.S. Deshpande and J. Prachita (2005), "WTO and Agricultural Policy in Karnataka.
2. Op.Cit.,
3. Report on Area, Production, Productivity and Prices for Crops, Government of Karnataka.
4. Op.Cit.,
5. Attri V.N. and Neera Varma, (2004), "Globalization, WTO and Sustainable Agriculture Development in India; Seminar on Institutional Policy Option for Sustainable Agriculture Development.
6. Misra and Lakshmi Kumari (2004), "Agriculture, Biotechnology and WTO: A case of AP., Seminar on Institutional Policy Option for Sustainable Agriculture Development.
7. R.S. Deshpande and J. Prachita (2005), "WTO and Agricultural Policy in Karnataka.

REFERENCES

Amitab Shukla (2000), "Regional Planning and Sustainable Development", Kanishka Publishers, New Delhi, p. 292.

Hanumantha Rao, C.H., L.K. Ray and K. Subbarao (1998), "Unstable Agriculture and Droughts, Vikas Publishing House Pvt. Ltd., New Delhi.

Parikh G.K. (ed) (1998), "Sustainable Development in Agriculture, Martinus Nijhoff, Amsterdam.

Srivastava S.P. (1998), "Development Debet: Critical Perspective", Rawat Publications, New Delhi, p. 320.

Surendar G. (2004), "Need of Balanced Fertilization in Agriculture for Sustainable Development", National Seminar on Institutional Policy Options for Sustainable Agricultural Development, Hyderabad, 10th and 11th April, pp. 160-165.

INDEX